my revision notes

D1335029

WJEC EDUQAS GCSE (9–1)
GEOGRAPHY B

Stuart Currie

HODDER
EDUCATION
AN HACHETTE UK COMPANY

The Publishers would like to thank the following for permission to reproduce copyright material.

Photo credits: p23 Adrian Sherratt/Alamy Stock Photo; **pp25, 30** Alyson Andrew; **p37***t* ZOOM DOSSO/ Stringer/Getty Images; **p37***m* U.S. Department of Defense Current Photos; **p37***b* FAYEZ NURELDINE/AFP/ Getty Images; **p38***l* WENN Ltd/Alamy Stock Photo; **p38***r* RAVEENDRAN/AFP/Getty Images; **p43** Janine Wiedel Photolibrary/Alamy Stock Photo; **pp52, 54***t*, **54***b*, **56** all, **58** Alyson Andrew; **pp65, 70** Ordnance Survey (licence number 100036470); **p71***t* Eric Foxley; **pp71***b*, **p77***l*, **p77***r* Alyson Andrew; **p86** Blaine Harrington III/Alamy Stock Photo; **p91** Tibor Bognar/Alamy Stock Photo; **p92***t* Vladyslav Makarov/123rf; **p92***bl* Phunrawin Nititadakul/123rf; **p92***br* goldfinch4ever/123rf; **p93 all** Alyson Andrew; **p96***l* Ethan Miller/Getty Images; **p96***m* blickwinkel/ Alamy Stock Photo; **p96***r* Sergey Utkin/Alamy Stock Photo; **p97** Images of Africa Photobank/Alamy Stock Photo; **p99** Kristian Buus/Alamy Stock Photo; **p101***l* Friedrich Stark/Alamy Stock Photo; **p101***r* Alyson Andrew; **p107** Paul Sampson/Travel/Alamy Stock Photo; **p111** Francesco Fiondella/Alamy Stock Photo; **p112** Joerg Boethling/Alamy Stock Photo; **p115***l* Eye Ubiquitous/Alamy Stock Photo; **p115***r* Joerg Boethling/Alamy Stock Photo; **p118** Ben Nelms/Bloomberg via Getty Images.

Acknowledgements: p23 Wales Online. **p27** Digital Strategy Consulting. **p31** Based on material adapted from the Lake District National Park's erosion factsheet. **p32** Figure 1 based on data from the World Economic Forum's Global Competitiveness Report. Figure 2 based on data from www.atkearney.com/research-studies/global-cities-index. **p40** OECD, The Organisation for Economic Co-operation and Development (OECD) – Our mission, www.oecd.org/about/. **p45** Figure 14 Human Development Report 2016 'Human development for Everyone'. Human Development Report Office, United Nations Development Programme. hdr.undp.org. **p47** Figure 16 Truewealth Publishing. **p49** The material in Figure 19 is adapted by the publisher from http://www.oxfam.org.uk/ what-we-do/emergency-response/east-africa-food-crisis [accessed June 2017] with the permission of Oxfam, Oxfam House, John Smith Drive, Cowley, Oxford OX4 2JY UK www.oxfam.org.uk. Oxfam does not necessarily endorse any text or activities that accompany the materials, nor has it approved the adapted text. **p51** Figure 1 BGS © NERC 2017. **p55** Urban Rim www.urbanrim.org.uk. **p84** Based on an image © UNFCCC. **p91** Substantially adapted from a table in 'The future of the urban environment and ecosystem services in the UK', Joe Ravetz, Centre for Urban Resilience and Energy, Manchester University, October 2015. **p95** *Science*. **p103** Figure 8 based on data from European Environment Agency (EEA). **p104** © Geological Survey of Denmark and Greenland (GEUS). **p105** John Sigerson/Executive Intelligence Review. **p107** 'New fresh water supply for Birmingham', from the *Birmingham Mail*.

Every effort has been made to trace all copyright holders, but if any have been inadvertently overlooked, the Publishers will be pleased to make the necessary arrangements at the first opportunity.

Although every effort has been made to ensure that website addresses are correct at time of going to press, Hodder Education cannot be held responsible for the content of any website mentioned in this book. It is sometimes possible to find a relocated web page by typing in the address of the home page for a website in the URL window of your browser.

Hachette UK's policy is to use papers that are natural, renewable and recyclable products and made from wood grown in sustainable forests. The logging and manufacturing processes are expected to conform to the environmental regulations of the country of origin.

Orders: please contact Bookpoint Ltd, 130 Park Drive, Milton Park, Abingdon, Oxon OX14 4SE. Telephone: (44) 01235 827720. Fax: (44) 01235 400401. Email education@bookpoint.co.uk Lines are open from 9 a.m. to 5 p.m., Monday to Saturday, with a 24-hour message answering service. You can also order through our website: www.hoddereducation.co.uk

ISBN: 978 1 4718 8737 6

Second Edition © Stuart Currie 2017

First published in 2013 by
Hodder Education,
An Hachette UK Company
Carmelite House, 50 Victoria Embankment
London EC4Y 0DZ

www.hoddereducation.co.uk

Impression number 10 9 8 7 6 5 4 3
Year   2021 2020 2019 2018

Cover photo © age fotostock/Alamy
Produced and typeset in Bembo by Gray Publishing, Tunbridge Wells, Kent
Printed in Spain

A catalogue record for this title is available from the British Library.

Get the most from this book

Everyone has to decide his or her own revision strategy, but it is essential to review your work, learn it and test your understanding. These Revision Notes will help you to do that in a planned way, topic by topic. Use this book as the cornerstone of your revision and don't hesitate to write in it – personalise your notes and check your progress by ticking off each section as you revise.

Tick to track your progress

Use the revision planner on pages 4 and 5 to plan your revision, topic by topic. Tick each box when you have:

- revised and understood a topic
- tested yourself
- practised the exam questions and gone online to check your answers.

You can also keep track of your revision by ticking off each topic heading in the book. You may find it helpful to add your own notes as you work through each topic.

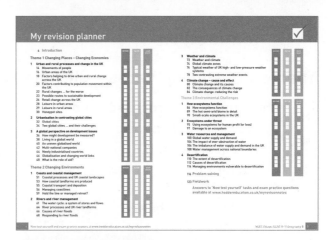

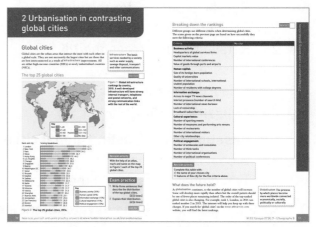

Features to help you succeed

Now test yourself

These short, knowledge-based questions provide the first step in testing your learning.

Definitions and key words

Clear, concise definitions of essential key terms are provided where they first appear. Key words from the specification are highlighted in bold throughout the book.

Revision activities

These activities will help you to understand each topic in an interactive way.

Exam practice

Practice exam questions are provided for each topic. Use them to consolidate your revision and practise your exam skills.

Online

Go online to check your answers to the Exam practice and 'Now test yourself' questions at **www.hoddereducation.co.uk/myrevisionnotes**

My revision planner

REVISED   TESTED   EXAM READY

REVISED   TESTED   EXAM READY

Answers to 'Now test yourself' tasks and Exam practice questions available at www.hoddereducation.co.uk/myrevisionnotes

Introduction

Why should I use this book?

These revision notes have been written to accompany the Eduqas GCSE (9–1) Geography Specification B course to help you get the best possible results in your examinations.

There is a great deal you need to know in order to obtain a result you will be proud of in this subject. You not only need to know the subject of geography well but, of equal importance, also need to know how to use all of your geographical abilities to get the best out of the examination experience. This involves understanding exactly what the examiner wants of you and being able to provide this in the examination situation.

But don't panic – that's where these notes come in! They take you back through all the main areas of content for your course and will also guide you in how to use this information to get the best out of your exams. By the time you have finished you will know almost as much about this as do your examiners – not a bad position to be in!

So, please don't ignore the opening pages of these notes. They are the key to getting the best out of the rest of the book and, as a result, obtaining the best possible result for *you* in geography. There are many candidates who are entered for the examination who are very good geographers but who never quite develop the ability to show this in the examination room. Read on and get involved in the activities to ensure that you are not one of these people.

Using these revision notes

These revision notes are not all you need to gain examination success. They have certainly not been written with the intention of replacing your teacher, the most important resource you have.

Most of you will have been studying geography since at least entering secondary school and will have learned a great deal in that time. You will probably also have notes in exercise books and files that will help you to prepare for your examinations. These notes and your teachers are the *real* key to your success in the examinations.

These revision notes will, therefore, help you to make sense of your own notes and help train you in the art of how to use your own geographical competencies to respond to the variety of tasks your examiners will set before you. Your teachers will also be working hard with you to ensure your examination success and it is my intention that these revision notes help in this process.

Unlike many revision guides, this book does not contain a huge amount of facts. These you already have. It does, however, help you to improve the abilities you will require to make the most of the examination experience; how to respond in the examination to the demands for you to apply your geographical knowledge, understanding and skills to new situations.

> **Exam tip**
>
> Are you naturally untidy? If so, train yourself in the art of keeping tidier notes. If your work is in exercise books, number these and write a brief content record on the inside front cover of each book. Create something similar for loose-leaf notes and don't forget to number the pages. If you don't do this, a ring binder depositing its contents on the floor could be very troublesome. Use this book as a summary of your work. Complete the activities provided and annotate the pages wherever you feel it will be helpful.

Getting to know your specification

It is easy to look on the examination and, sometimes, your school work, as the enemy. It really shouldn't be that way. This specification was written so that your teachers can create a course for you that is both enjoyable and relevant. The examinations are also designed to help you.

The Eduqas GCSE (9–1) Geography Specification B

The following is an overview of the three Themes that make up the Eduqas GCSE (9–1) Geography Specification B. The first key to doing well in your examinations is to develop an understanding of each of these.

Theme 1: Changing Places – Changing Economies	Theme 2: Changing Environments	Theme 3: Environmental Challenges
Urbanisation in contrasting global cities	Coasts and coastal management	How ecosystems function
Urban and rural processes and change in the UK	Rivers and river management	Ecosystems under threat
Global perspectives on development issues	Weather and climate	Water resources and management
	Climate change – cause and effect	Desertification

Assessment through Eduqas GCSE (9–1) Geography Specification B

The structure of the entire examination (the big picture) is shown in the table below. All components are written papers and all the questions asked in them are compulsory.

Name	Nature of assessment	Time	Marks/proportion of total mark
Component 1	Three questions: one for the content of each of the three themes	1 hour 45 minutes	96 marks/40%
Component 2	A three-part paper introducing you to a geographical problem, exploring different solutions and asking you to justify your choice of solution	1 hour 30 minutes	72 marks/30%
Component 3	Also in three parts. The first two parts test your ability to apply your fieldwork techniques to new situations. The third part tests your understanding of wider UK concepts and to make a UK-based decision	1 hour 30 minutes	72 marks/30%

The examination ... in greater detail

You will sit all of the papers that make up your examination in three sessions known as Components. They are usually a few days apart and in June of your Year 11.

The **first Component** will test all of the three Themes you have studied. These are:
- Changing Places – Changing Economies
- Changing Environments
- Environmental Challenges.

Most of the questions you will answer will be based on resources like maps, graphs, diagrams and photographs that are a part of the paper. At least one question will explore your knowledge and understanding of the UK as a whole.

The **second Component** will test your ability to solve a geographical problem. The geographical content of this component will be taken from across all three of the Themes and, as suggested by its name, will lead you through three 'Parts':
- The first will introduce you to a place and a problem that needs to be solved there.
- The second will ask you to explore different solutions to the problem.
- The third and final part will ask you either to select one of the solutions offered or to rank those offered in order of importance. You will then need to justify the choices you have made.

There is a chapter about 'problem solving' later in this book.

The **third Component** is in three parts. The first two parts are linked and will test the techniques you have learned when carrying out two fieldwork enquiries; one investigating an aspect of human geography and the other, physical geography. These are called **fieldwork methodologies**. There are four of these, *Use of transects*, *Change over time*, *Qualitative surveys* and *Geographical flows*. You will be tested on only one fieldwork methodology. Find out which one it is and add it to the box below.

It will also assess your ability to use fieldwork findings to explore one 'conceptual framework'. There are six **conceptual frameworks**: *Place*, *Spheres of influence*, *Cycles and flows*, *Mitigating risk*, *Sustainability* and *Inequality*. You will only be tested on one conceptual framework in your examination. Find out which one you will be tested on and add it to the boxed sentence below.

The final part of the third component tests your ability to make a decision based on your wider understanding of the geography of the UK.

> In Component 3 I will be tested on:
>
> my ability to use _____
>
> _____ fieldwork methodology
> *and* my ability to use fieldwork findings to
> explore the _____
>
> _____ conceptual framework.

Now test yourself

TESTED

Add geographical terms for each of the phrases in the table below. Other key terms are highlighted throughout this book. Get used to using them when answering questions.

Phrase	Term
Wearing away ... of rocks	
Shopping bought on a near-daily basis	
Jobs with a regular wage and tax paid	
Mining on the Earth's surface	
Growing enough food for just the family	

Phrase	Term
A small river that flows into a larger one	
Planting a large area with trees	
Growth of the proportion of people living in towns and cities	
Reasons why people move away from their home area	

SPaG

Teachers often talk about SPaG: it stands for 'spelling, punctuation and grammar'; in other words, the quality of your written English. It also takes into account your ability to use geographical terms effectively.

There are four marks available on each of the three Components for your SPaG abilities. They will be awarded for your responses to those questions that have the most lengthy spaces for answers.

Making the exams work for *you*

How well you do in your exams is mainly down to you. You are the only one who can give your studies enough time throughout the course and to ask for help if you are struggling at any time. However, you are just the most important point of a triangle that involves other people.

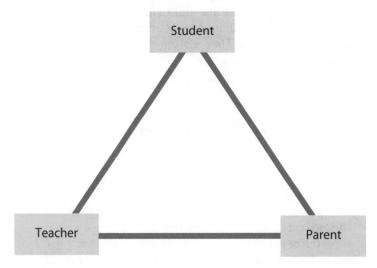

Are there any other people you could add? How might they help you?

There is certainly one other group. It's worth shouting out loud, '*your examiners really want you to do well*'. They use quite a few different ways to make sure that you provide them with the answers that they want from you. This can only happen, though, if you play their game. They expect you to:

1 Read all the information provided on the examination paper. There may be no questions to answer on its front page, for example, but there are instructions and other information to help you.

2 Manage your time carefully. You must actually complete the paper if you are to do well. Your examiners help you by telling you how many marks are available for each question and by giving you the number of lines they feel you will need for your answer. In Component 2, the problem-solving paper, they even advise you on how much time to spend on each of its three 'Parts'.

3 Do everything that is asked of you. Your examiners use 'command terms' to tell you what sort of answer they want from you. When used, each of these terms always means the same thing so you must always respond to them in the same way. Practise this throughout your course.

Understanding the command terms

The questions your examiners are going to ask fit into four different groups. The examiners call these Assessment Objectives (AOs). On the front of each of your examinations there will be a reminder of what each AO means and every question you answer will inform you of which AO it is testing. These are shown in the table below.

Type of question	Command terms	How to respond	Can do (✔)
AO1 You can show your knowledge of locations, processes, environments and different scales	list, name, give, circle underline, recall 1/2/3 facts about	Do no more than you are asked. Don't even attempt to describe the feature/fact you've been asked to name or recall	
	give the meaning of, describe how	Extend and elaborate beyond just naming something. Give as many different pieces of description as there are marks for the question. Don't, though, explain why something happens. That is often the answer to the next question	
AO2 You can show understanding of concepts related to, and interrelationships between places, environments and processes	Give one reason, give reasons to agree/disagree with … Explain why	This is your opportunity to show that you understand why something is happening. The word 'because' is worth using in your answer to make sure you keep on track	
AO3 (1) Analysis: you can use your knowledge and understanding to analyse and interpret information provided in the paper	Use information from the graph to suggest how people's lives will be affected by … Use map evidence to suggest why …	This goes a step further than the description of resources that are asked for in AO4. You are expected to use your findings in order to bring geographical sense or meaning to them. For example, to suggest ways in which the information may affect people or the built and natural environments	
AO3 (2) Evaluation: you are asked to use evidence to decide how good a particular activity, plan or strategy is, whether in isolation or compared to others	Weigh up the strengths and weaknesses of … Explain why *x* strategy is more sustainable than *y* strategy	Now is the chance for you to start giving your own opinions on a particular issue You are asked not only to compare but also to make a value judgement about the options being offered. Always back up your judgements with evidence either from a given resource or from your own geographical knowledge	
AO3 (3) Making judgements: you will need to use your *evaluation* of a situation to decide the best way forward to a sustainable future	Choose the best/most sustainable strategy … Prioritise/rank in order of importance (3) strategies … … and justify your choice	It's a very short step from evaluation to here. Your evaluation of any alternative strategies to solve a geographical problem or issue will have triggered preferences in your mind You will need to make a decision based on these preferences and justify it	

Type of question	Command terms	How to respond	Can do (✔)
AO4 You are able to use your skills and techniques to both complete and read resources like maps, graphs and tables, and to state what you have found	Complete the graph using the following figures Describe the pattern ... on the map Work out the mean ... Compare the changes in electricity production in country *X* and country *Y*	Read the question carefully. It may ask you to describe, for example, simply what is happening along a single line graph or it could ask for a comparison between the lines for two different places. In this case the term 'whereas' is often a useful one to use *Do* quote clear evidence from the source, often figures *Don't*, though, go further. The question is not asking you for a judgement as to which is better for a particular purpose	

Finally, your examiners sometimes underline or print in bold some words in a question if they want to draw your attention to them. There is no reason why you should not do the same to help you understand exactly what is expected of you.

Now test yourself

TESTED

For each question below:
1 underline the key words that you feel help you to understand the question
2 decide what type of question it is, based on the 'type of question' column in the table above
3 draw a brief outline of your response to the question.

a) Explain why government in the UK may wish to reduce the threat of climate change. [6]
b) Many central business districts (CBDs) have pedestrianised areas. Describe the features of pedestrianised areas. [2]
c) Evaluate the use of the internet as a source of secondary data to support fieldwork. You should support your answer by referring to actual examples from your own fieldwork. [6]
d) Name an ecosystem you have studied. For your chosen ecosystem, complete the table below to give the names of specific plants or animals found in your chosen ecosystem. [3]

	Plants/animals in ecosystem
One tertiary consumer	
One secondary consumer	
One primary consumer	
One producer	

Effective revision

The only person who can decide how to revise most effectively is *you*. There are a huge variety of techniques and some will suit some people more than others. Perhaps the following quiz will help you to decide your most effective means.

Question	Yes	No
Do you need complete silence to revise?		
Does music help to cut out outside noises?		
Can you concentrate for long periods of time?		
Is your attention span short?		

- For how many different exams do you have to revise?
- Where is geography in your exam timetable?
- What parts of your social life are essential?
- What can you give up to make time for revision?

There is just one rule when answering these questions – be totally honest with yourself. There is a very long period of time between your examinations and the results day. You will only enjoy this time fully if you have completely prepared yourself for the examinations and you can honestly say that you could not have done better.

So, now you are armed with this vital information, create your own customised revision programme that:
- starts early enough
- balances work and pleasure
- suits *your* concentration span
- is realistic in the demands it places on *you*
- takes place in conditions that suit *you*
- builds in rewards.

Finally, remember just three more points:
- Your teacher is there to help and will welcome questions.
- There may be geography lessons offered during any study leave time you may be given – attend them!
- Everyone realises the pressure you are under. If you feel, at any time, you are not coping, be sure to talk to someone about it.

Active revision

Your revision can be either *active* or *passive*. Passive revision involves just reading your notes and is something that is likely only to work over very short periods of time. After this the mind begins to wander and all sorts of outside influences will get in the way of effective revision: things like staring at posters on your wall or listening for noise coming from outside your room.

On the other hand, *active* revision involves you in actually *doing* something. This action is likely to help you to maintain your concentration at a reasonably high level and can often result in you also producing something that will be helpful later in the revision process.

There are a number of activities you might attempt:

- Make revision cards for the main issues you have studied through the course. The breakdown of Themes on pages 4 and 5 of this book suggests titles that you may wish to use for your cards. Include basic detail of the factual content you hope to use in the examination.
- Link features affecting or influencing a particular feature by drawing spider diagrams or webs. This is useful for such ideas as influences on quality of life of, for example, local service provision or the effects of flooding on an area of a river flood.
- Create cards to test yourself and your friends on some of the key terms needed for success in a geography examination. Produce one set of cards of the terms and another set with their definitions. Use them as a simple matching exercise or a game of geography 'snap'. You might consider creating a set for each of the three Themes.
- Well-drawn sketch maps and diagrams are always welcomed by your examiners. Try to memorise some of these and then attempt to redraw them. Compare your redrawn maps with the originals. Go a little further and adapt these to match exactly a question being asked.

Each of these strategies for revision success is looked at in greater detail as you work through this book.

And finally …

The big day has come, you have revised well and there is nothing that can get in the way of your success. Or is there?

Meeting these checklists will further help you to succeed by keeping you in control of the situation. Complete this for your geography exam.

You will:

reach the exam room in plenty of time	**Component 1** date: time:
	Component 2 date: time:
	Component 3 date: time:
know your centre and candidate number	Centre: Candidate:

You will also:
- listen carefully to your invigilators
- carry spare writing equipment
- read the front-page instructions
- use your time well
- answer all questions – don't leave gaps.

Exam tip

- Discuss the points in this checklist with friends, teachers and parents.
- How will each point help you stay in control?
- Are there any other ideas that will help you?

1 Urban and rural processes and change in the UK

Movements of people

Populations are dynamic; they are constantly changing. This is due to:

- **Natural population change**: natural increase (when there are more births than deaths each year) tends to be greatest in **low-income countries (LICs)** and **newly industrialised countries (NICs)**.
- **Migration**: people migrate into and out of both the countryside and towns and cities. There are many reasons for these movements of population. They change with time and differ from place to place.

Revision activity

Complete the following table to give an example of each type of migration. Include at least one further example of temporary migration. You may wish to draw sketch maps to show these movements.

Type of migration	Example of movement from/ to which places	Reason for movement	Permanent or temporary?
Rural to urban			
Recent international	Australia to London	Many young Australians are attracted by London's entertainment and lifestyle. They also use it as a base for visiting other EU countries. They usually return to Australia within a year	Temporary
Urban to rural			
Historic international			
Region to region	Various Indian cities to Bengaluru	In the early twenty-first century, many graduates have been tempted to work in the rapidly growing technology industry. It offers high wages and pensions	Permanent
Temporary national region to region			

Natural population change Change in the number of people living in a place as a result of differences in births and deaths

Low-income countries (LICs) Countries which had a gross national income per capita of $1025 or less in 2015. This figure changes yearly

Newly industrialised countries (NICs) Countries which in recent years have greatly increased their manufacturing capacity

Migration The movement of people to live in a different place. Migrations may be permanent or temporary

Rural to urban migration Migration from the countryside to towns and cities

Regional migration Movement from one region to another in the same country

International migration Migration from one country to another

Involuntary migrations

REVISED

You may have noticed that, in the table on page 14, not all migrations are voluntary. In some cases the migrants have no option but to move. The Syrian civil war that started in 2011 had resulted in the migration of over 13 million people from their home settlements by 2016. 6.6 million are displaced within Syria. Another 4.4 million are living in the neighbouring countries of Turkey, Jordan and Lebanon. About 13.5 per cent of the total have travelled to the EU.

Economic migrants Move out of choice. Usually attracted by the prospect of a better job and living conditions

Refugees Move because they are forced away from a place where their lives are in danger

Asylum seekers Refugees in fear of persecution in their country of origin for reasons of political opinion, religion, ethnicity, race/nationality or membership of a particular group

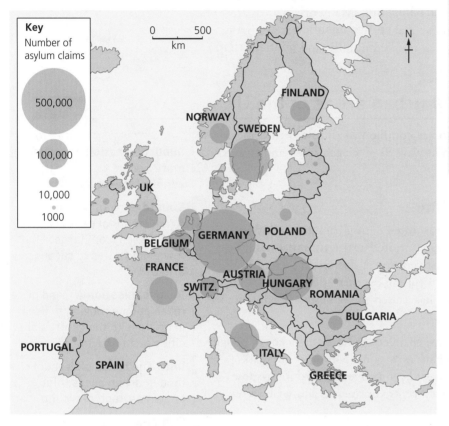

Figure 1 Asylum claims in Europe, 2015.

Now test yourself

TESTED

Figure 1 is a proportional circle map. Complete the table below to suggest two advantages and two disadvantages of showing information in this way.

Advantages	Disadvantages
Gives an immediate impression of distribution	

Exam practice

Study the map in Figure 1.
a) Describe the distribution of asylum seeker applications in Europe during 2015. [3] [5 lines]
b) Explain why asylum seekers moving into an area might put pressure on a city's services. [3] [5 lines]

ONLINE

Urban areas of the UK

Urban changes in the UK

REVISED

Historically, UK urban areas have grown because of the need for people to be close to places of work. Much of the work was in factories. These caused air pollution, and **suburbanisation** took place as those people who could afford to moved out to the edges or as new council-owned houses were built for rent. Improved public transport and car ownership helped this and also encouraged **counter-urbanisation**. Many urban areas are now much more pleasant areas to live and **re-urbanisation** is occurring. A shortage of houses to meet this has resulted in **infill** and **gentrification** taking place.

Distinctive features of urban areas of the UK

REVISED

There are a number of factors that are common to all towns and cities. However, they combine in different ways to give each town and city its individual character.

Physical changes over time

The historical growth of urban areas, often dating back to the Industrial Revolution of the nineteenth century, created a structure that has little relevance to lives in the twenty-first century. A decline in secondary industry in the second half of the twentieth century and an increase in the ability of most people to travel has caused the old established structure to be adapted.

As the functions of urban areas have changed, so have the needs of people who live in them. Most factories have now gone. More people work in offices. Many inner-city areas have been converted into high-value housing for people who work in city-centre businesses and wish to travel to work on foot.

Economic changes over time

Where people live usually depends on their ability to either buy or rent a property. The more desirable areas tend to be more costly and attract people with higher earnings. In the past, there was a general increase in house values the further out from the centre you moved. Inner-city infill is now challenging the suburbs as high-priced housing areas.

Social changes over time

There is a long history of people from the same cultural background grouping together. During the Industrial Revolution one such group was Irish immigrants. Today, there are enclaves of people from some Commonwealth countries and countries in the EU. While there is some movement out of these areas as families become economically successful, there is also a tendency to remain in the community despite such success.

The complex changes that have taken place have resulted in different zones forming in urban areas. These are defined in the boxes on page 17.

Suburbanisation A trend for more people to live on the edges of towns and cities

Counter-urbanisation A movement of people and businesses from large towns and cities to rural areas

Re-urbanisation A trend of more people moving to live in or close to the centres of cities and large towns

Infill The reuse of derelict land in urban areas. Also known as 'land recycling'

Gentrification The conversion of existing buildings into high-quality dwellings

Urbanisation The physical and human growth of towns and cities

---

Characteristics of zones or areas in a town or city I have studied

Revision activity

a) Draw a sketch map of the town or city on a separate sheet of paper.
b) Label areas of your map to show the locations of each of the zones shown below.
c) Choose from the separate box of descriptions, a description for each of your zones.
d) Write three brief statements in each box to describe the zone's characteristics

A box below has been left blank for you to add a zone found in your settlement that isn't already named.

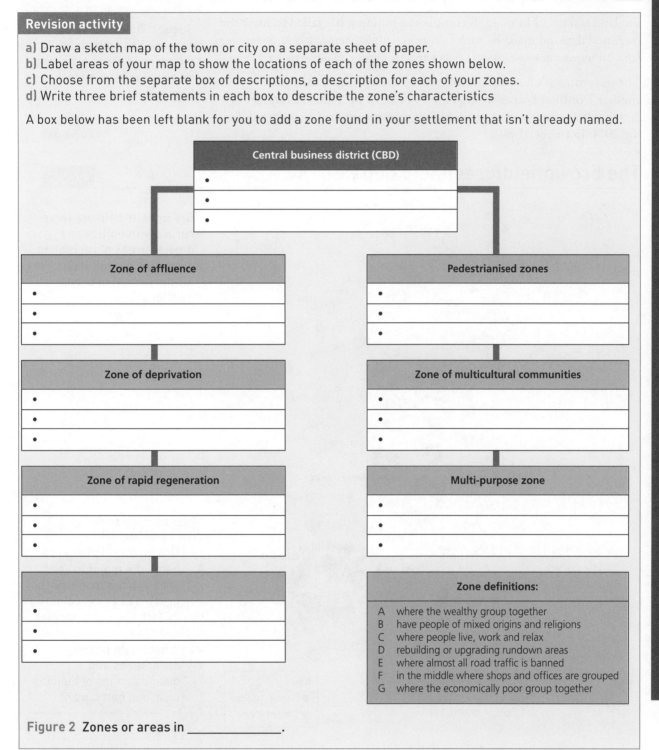

Central business district (CBD)
-
-
-

Zone of affluence
-
-
-

Pedestrianised zones
-
-
-

Zone of deprivation
-
-
-

Zone of multicultural communities
-
-
-

Zone of rapid regeneration
-
-
-

Multi-purpose zone
-
-
-

-
-
-

Zone definitions:

A where the wealthy group together
B have people of mixed origins and religions
C where people live, work and relax
D rebuilding or upgrading rundown areas
E where almost all road traffic is banned
F in the middle where shops and offices are grouped
G where the economically poor group together

Figure 2 **Zones or areas in _____.**

Factors helping to drive urban and rural change across the UK

England is short of housing. Recent house building has failed to meet the increased demand made by such factors as people *living longer* and *marrying later*, an *increase in single-parent families* and an *increase in immigration*.

The government's housing minister stated in September 2015 that another 1 million houses would have to be built by 2020. This target of 200,000 houses per year is much higher than the under 125,000 built in the 2014/15 financial year.

The brownfield/greenfield debate

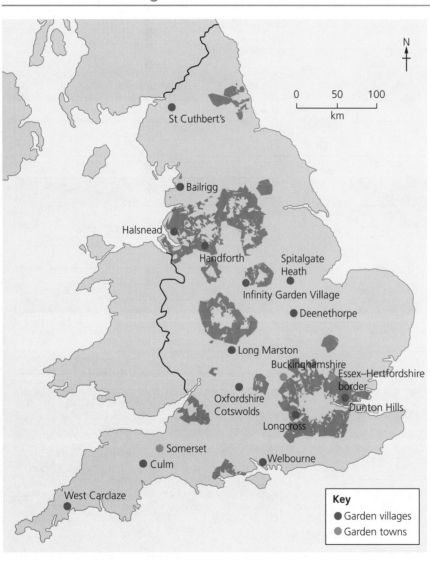

Figure 3 Green belts and planned developments.

Following the Second World War, large numbers of houses were built, resulting in towns and cities growing outwards into the countryside. **Green belts** were created to stop this expansion. Now that more housing is required there have been proposals to build new **garden towns** and garden villages to help solve the problem. Some of this building is planned for green belt land.

Making and justifying a decision: what would you do?

Complete the table below:

- Each statement is a simple fact about the site. You need to elaborate on it as an advantage or disadvantage of developing the site. Use specific information from your studies to help you. You could write your advantages in green and disadvantages in red. The first one has been done for you.
- Finally, complete the right-hand column to state whether it is a social (S), economic (E) or environmental (En) effect. Suggest another statement of your own for each type of site.

> **Brownfield site** Land suitable for redevelopment. Usually in urban areas
>
> **Greenfield site** Land previously unused for building. Usually in rural areas

Site	Statement	Elaboration
Brownfield site	Mainly areas of unsightly disused or derelict land	Improves the visual appearance of the area (En)
	Existing buildings can be adapted to housing	
	Increases demand on existing public transport	
	Reduces commuting distance	
	Increases cars on city roads	
	Reduces urban expansion	
	Utilities like water and power are already in place	
Greenfield site	No existing buildings to clear away	
	May change the character of the area for existing residents	
	Building not constrained by limited space	
	Could reduce farmland	
	Slow to get planning permission	
	Potential damage to habitats	
	Land unlikely to have been polluted by a previous use	
	Increases overall use of cars	

Revision activity

Now make the decision:

- Where would you recommend future housing development to take place? On brownfield or greenfield sites or a combination of both?
- Fully justify the decision you have made by referring to both types of site and including examples from your studies.

Factors contributing to population movement within the UK

Regional variations

There are many differences between regions within the UK. The south-east of England, for example, is traditionally regarded as being the most wealthy UK region, with people becoming poorer as we travel north and west. This is a very simplistic picture, though.

Looking at average earnings doesn't always tell you a great deal about real wealth as areas of high earnings may also be expensive areas in which to live. Higher transport costs, **council tax** and **mortgages** or rent payments could leave people with similar or lower **disposable incomes**. And, of course, averages for any region hide the differences that affect the lives of individuals living in these regions.

These can affect movement between regions. People attempting to move to the south-east for a job may be put off by the high cost of living, especially the cost of renting or buying a house. On the other hand, people who live in the south-east may have the opportunity to sell their house and buy one more cheaply as a retirement home in an area with a lower cost of living.

> **Council tax** A tax placed on each property that is payable to the local council
>
> **Mortgage** The repayment to a lender, like a building society, of money borrowed to buy a property
>
> **Disposable income** Money left after all essential payments have been made

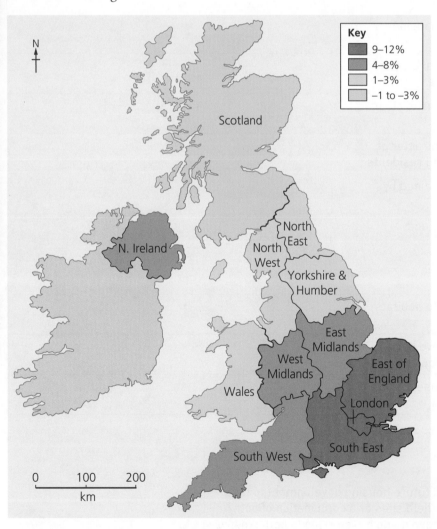

Figure 4 UK house price changes 2015–16.

Key
- 9–12%
- 4–8%
- 1–3%
- −1 to −3%

Exam practice

Study the map in Figure 4.
a) Describe the pattern shown on the map. [3] [5 lines]
b) Suggest why the ability to migrate between UK regions depends on where a person lives. [3] [5 lines]

ONLINE

Commuting and teleworking

The daily commute

In 2011, over 60 per cent of UK workers used a car to **commute** to work. Others travelled by public transport (train, bus, tram) or on a bike or by walking. The decision of how to travel has implications for the sustainability of the UK's urban areas. The more cars travelling into a town or city, the higher the air and noise pollution and the greater the congestion on its roads.

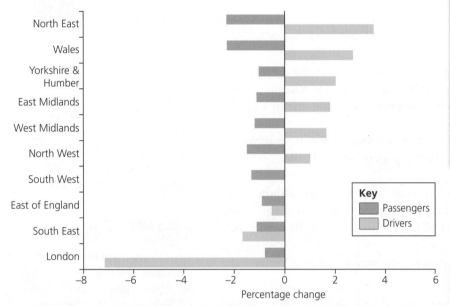

Figure 5 **The changes in car use for commuting across England and Wales, comparing 2011 with 2001.**

Teleworking: an alternative to commuting?

In 2015, 7.4 per cent of the UK workforce spent at least one day working from home. These employees link with work using their computers. Continued improvements in internet connectivity should encourage the present upward trend to continue. Almost two-thirds of the total were from the 'professional and managerial' **socio-economic group**. Some people 'telework' all of the time. The only influence of their work choices on where they live is the availability of a reliable internet connection.

	Advantages of working from home	Disadvantages of working from home
To employer	Office costs reduced Not disrupted by traffic problems	Less control of staff Difficulty linking team members
To employee	Work in comfort of home Lower travel costs	Difficult to separate family and home life Temptation not to work

The popularity of **telecommuting** could have major positive environmental effects, resulting in the development of more **sustainable** urban areas. Consider the effects of this on air quality and transport movement.

Commute Daily travel to and from home to a place of employment

Socio-economic group A way of categorising people according to their employment

Telecommuting Working part of the week in an office and the rest at home linking with the office by computer

Sustainable Capable of being able to operate effectively now and in the future

Exam practice

Study both the graph in Figure 5 and map in Figure 4.

a) Compare the pattern of changes in car use with that of house price trends. [3] [5 lines]

b) Suggest two reasons for the differences in changes in car use between London and Wales. [4] [6 lines]

ONLINE

Now test yourself

1 Suggest which, the map or the graph, is the more useful resource for predicting longer-term changes. Explain your choice.

2 How might increased teleworking result in more sustainable urban areas?

TESTED

Rural changes ... for the worse

Populations are dynamic. They are constantly changing. **Pull factors** encourage people to move into rural areas. They will have both positive and negative effects on the areas they move to. Many international migrants have moved into rural areas both on a permanent and a temporary basis. They form a labour force that harvests crops and keeps the rural economy going.

Commuter settlements

REVISED

- Rural areas are often thought of as *quiet* with *little air pollution*, as being *friendly* and *crime-free*.
- This encourages the migration of people outwards to the countryside. In some cases this is to buy second homes for use at weekends and for holidays. Other people sell their urban home and buy one in a rural area for retirement.
- Many working people have also left urban areas. They have moved to live in surrounding villages, turning them into **commuter settlements**.
- As these people work in the urban areas, they still tend to shop there, use their cars to commute and to drop off their children at school on their way to work, and be attracted to the nightlife of the urban area they have left.
- Services within and to the commuter settlements often suffer and are forced to close down.

Not all rural communities are sustainable. People find that they cannot afford to live in the area or that it does not give them the social lives they want. These **push factors** are likely to cause them to migrate away from the area. Below are the views of two Welsh MPs worried about more isolated rural communities.

> **Pull factors** Perceptions of a place that attract people to it

> **Commuter settlement** A place that has a large proportion of commuters
>
> **Push factors** Negative features of a place that influence people to leave it

> **Revision activity**
>
> 1 Create a spider diagram using the pull factors shown in *italics* in the text.
> 2 Add views of urban life that you have seen in the news on TV.
> 3 Add reasons why each would encourage people to migrate away from the urban area to a commuter settlement.

> Rural families are spending £2700 a year more on everyday goods than their urban counterparts

> Those who live in a rural area are likely to travel 10,000 miles a year, whereas those in urban areas travel 6400 miles

> Will someone who has a young family go and live in a village if the school, post office, pub or shop has shut, and if public transport is minimal?

> When an area loses many people, capital grants are reduced and that makes it even more difficult to sustain and regenerate local communities

> When walking into a house, prospective buyers look at their mobile phones to see how good the reception is and then ask what the broadband provision is like

> **Revision activity**
>
> What is meant by a sustainable rural community? List the factors shown on this page that would suggest that many rural communities are unsustainable.

Possible routes to sustainable development

Sustainable rural areas: villages fighting back?

REVISED ☐

One of the main constraints stopping many rural communities being sustainable is the lack of an effective bus service. Older members of communities often cannot drive and feel isolated. Local shops and services have closed and it is difficult to reach those in the urban areas. There seems to be little public money to provide these necessary services but there are charities that may help.

Llandysul is a market town 15 miles south of New Quay in Ceredigion.

Welsh 'rural riders' bus service tackles rural issues

At last a bus service aimed at making life easier for an isolated Welsh community has been given a chance to change people's lives. A Llandysul-based charity has been awarded almost £5000 by the Big Lottery to set up a new bus service.

The single bus will link the town with residents who live in surrounding villages. They will hop on the bus and be taken to the centre of the market town.

There are about 50 different community groups and schools that will also use the vehicle.

The project will be run by a not-for-profit company that aims to promote the area's economy, environment, culture and the quality of its community life.

The Plunkett Foundation

The Plunkett Foundation is a charity that supports local communities in their attempts to keep local services like shops and pubs open. Funding is dependent on the local community also investing in the service by buying shares in it.

The Avebury Community Shop, Wiltshire, is supported by the National Trust, which provides and maintains the shop building at a low rent, and the Plunkett Foundation, which provides ongoing financial support. It was opened in March 2009 and, run by mainly volunteers, continues to serve the village.

> **Revision activity**
>
> Use information from these pages to suggest what might be done to make rural communities sustainable. Create a table to help you organise the strategies according to whether they would affect social, economic or environmental sustainability.

Post office services

Café area

Local farm produce

Displays of local art and jewellery

Household essentials

Meeting room

Newspapers

Photocopying

Figure 6 Customers at the community-run shop in the Wiltshire village of Avebury.

A sustainable urban community?

What does 'sustainable' mean?

To be sustainable, a town or city must go some way towards meeting the qualities shown in the list below. If we were to build completely new settlements like garden towns (see page 18), they would be constructed to be sustainable. However, most people live in settlements that have developed over centuries and the desired twenty-first-century sustainability must be gained by adapting what's already there. Most urban areas are planning for increased sustainability through a mix of public and private initiatives.

Some sustainability criteria

Criteria	Possible constraints	Nottingham
Resources and services are accessible to all	May be difficult to reach due to distance or to access according to disability. Language may also be a difficulty	
There is access to affordable housing	Mortgages may be difficult to find. Houses of the right size or price may not be available to all groups of people	Nottingham Community Housing Association, a 'not for profit' group, aims to provide homes to rent for those in need
Different communities work together	There can be a tendency for people of a particular culture, income group or religion to live near one another. Chances for mixing may be few	The free three-day annual 'Riverside Festival' is a celebration organised by the city council for and run by the whole community
Cultural and social amenities are open to all	Even when some places are physically accessible, there may still be constraints based on such factors as income and gender	Nottingham Playhouse: 'Everyone is welcome to attend our Pay What You Can Performances and pay whatever they can. 1p, 50p, £1, £5'
There is investment in the central business district (CBD)	CBDs have become less popular with retailers and service providers due to changed commuting habits and increased internet sales	Construction of a new £30 million bioscience hub completed in March 2017. It supports more than 300 jobs
Public transport is good enough to consider as an alternative to driving	Access from the home to a destination often involves a long walk and a change of bus or train	
Walking and cycling are both safe activities	It's difficult to add 'safe' cycle lanes in particular to existing city roads. Roads are often too narrow to add footpaths	£6.1 million to be spent on cycle 'super-highways' to link Beeston, Clifton and Hucknall to the city centre, They will be physically separated from roads

Exam practice

'It is only possible to create a sustainable community by improving its transport system.' To what extent do you agree with this statement? Make use of evidence from these pages in your answer.   [8] [30 lines]

ONLINE

Revision activity

1 Use information from the sketch map in Figure 7 to complete the table for Nottingham.
2 Create a third column for another town or city in an HIC you have studied.

The Nottingham Express Transit (NET) network

NET is the Nottingham city tram system. The first development, from the city centre to Hucknall, was opened in 2004. The other routes started operating in 2015. A GCSE student collected the evidence for the land-use map on this page during a journey from the Toton Lane tram terminus to Nottingham railway station.

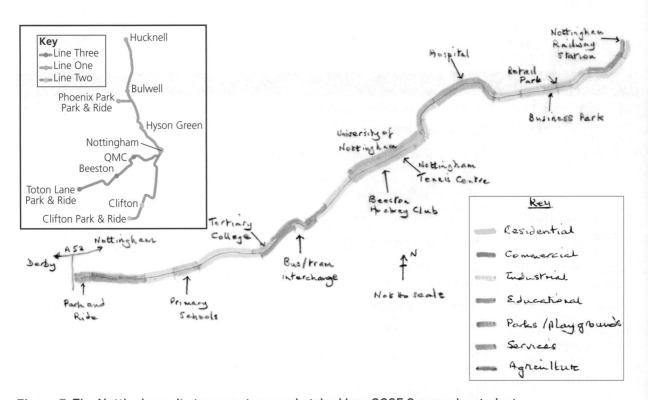

Figure 7 **The Nottingham city tram system as sketched by a GCSE Geography student.**

Figure 8 **Access to the Queen's Medical Centre (QMC) from its tram stop.**

Revision activity

The industrial zone next to Nottingham Tennis Centre is the Nottingham Science Park. A science park is a place where the industry taking place is based on scientific research.

1 Locate and name the Nottingham Science Park on the sketch map in Figure 7.
2 Why might this be a good location for a science park?

Retail change across the UK

Shopping patterns are constantly changing. In recent years we have seen the decline of town- and city-centre shopping centres along with the rise in out-of-town retail parks. This has been mainly the result of the growth of access to private transport. This allows customers to readily access such centres. As inner-city areas have been redeveloped, most corner shops have disappeared. What are the characteristics of the main types of shopping experiences?

High-order goods Expensive items, bought infrequently

Low-order goods Household items, bought frequently

Catchment area The area around a shop/shopping centre from which it draws its customers

The main types of shopping experiences

REVISED

Type and example	Characteristics	Advantages	Disadvantages
CBD: in the middle of large urban areas, for example	High- and low-order goods Banks and offices Central to public transport and roads Large **catchment area**, **threshold population** and **range**		
District shopping centre: in centres of small towns and suburbs of cities, for example	Smaller variety of shops Mainly low-order goods Smaller catchment area, threshold population and range High percentage of vacant shops		
Malls: group of small stores under one roof. Often in town centres, for example	Often include entertainment Have a variety of high- and low-order goods Dedicated car parks		
Retail parks: usually located out of town, for example	A variety of high- and low-order goods Close to main roads Dedicated car parks Large catchment area		
Corner shops: usually on ends of terraced housing rows, for example	Low-order goods Low catchment area, threshold population and range Accessed on foot Open long hours		

Now test yourself

TESTED

1 Add an example of each shopping centre to the table above.
2 Complete the advantages and disadvantages columns for each shopping centre. Use those in the list below and add some more from your own studies. You could use the same feature more than once: *space to develop, congested, free parking, expensive parking, difficult access for elderly, dominated by charity shops, cramped location, all indoors, rents expensive.*

Threshold population The number of shoppers required to keep a shop in business

Range The distance a consumer is willing to travel to buy a particular product

The technological revolution

The way we conduct our everyday lives has changed greatly during the early part of the twenty-first century … and is still changing. One of the main areas we see this is in the retail industry. A change that appears to have failed is that of 24-hour shopping. Many large supermarkets abandoned this in 2016 as being too expensive at a time of supermarket price wars. However, it is the use of the computer for many of the purchases previously made in the stores that has changed everyday shopping habits.

	2013	2014	2015	2016	Increase 2014–16
UK	£38.84	£44.97	£52.25	£60.25	15.3%
Germany	£28.98	£36.23	£44.61	£54.60	22.4%
Sweden	£3.13	£3.61	£4.17	£4.87	16.8%
EU	£111.23	£132.05	£156.67	£185.44	18.7%
USA	£165.30	£189.26	£215.38	£245.96	14.2%
Canada	£11.81	£12.84	£14.53	£16.65	14.5%

Figure 9 **Changes in internet shopping (figures are in billions).**

Now test yourself

TESTED

You are asked to show on one graph the UK data in the table above compared to that of two other countries.
1 Which other two countries would you choose? Why?
2 What information would you show? Explain your choice(s).
3 What graph type would you use? Why?

Death of the high street?

In January 2017, Parcelhero, an international delivery group, made a number of predictions about changes by 2030. They included:
● almost half of all high street stores will close
● e-commerce will be 40 per cent of all retail sales
● the e-fashion market will rise from 21 to 63 per cent.

Some retailers have already responded to the threat by developing online sales themselves. Tesco's £2.9 billion internet revenue is second to Amazon UK. John Lewis already has 25 per cent of its total sales via the internet. The suggestion is that town centres will have more housing and concentrate on the entertainment and restaurant trades.

Revision activity

Make a list of the possible effects of increased internet shopping. Divide it into social, economic and environmental effects.

Leisure in urban areas

Leisure facilities are spread throughout our towns and cities. There are, though, concentrations in certain locations:

● Traditionally, the CBD has a high proportion of an area's theatres, cinemas, clubs, restaurants and pubs. This has increased as CBDs and their immediate surrounds have been regenerated.
● There are also concentrations in places on the edges of towns and cities that are readily accessible by car. Food halls and entertainment are major features of out-of-town shopping centres, for example.

Advantages and disadvantages of leisure use

Such businesses create advantages in providing employment:

● **Direct employment** like in theatre box offices or restaurants' chefs and waiters.
● **Indirect employment** by support industries like transport and waste disposal.

On the negative side, there are potential problems, especially at night in areas that are also residential. Noise and anti-social behaviour, often resulting from alcohol consumption, are especially an issue at weekends.

> **Direct employment** Jobs created within a particular business
>
> **Indirect employment** Jobs created outside a business but that depend on it for their existence
>
> **Inertia** An inability to move because of the high cost of relocation

Exam practice

a) List three night-time problems for city-centre residents. [3] [3 lines]
b) Explain why residents may still wish to live in these areas. [3] [5 lines]

ONLINE

Major sporting events

These have different characteristics to the leisure activities described above. The stadiums in which they take place are often in residential areas, sometimes because they were built there many years ago and cannot move because of **inertia**. Others have recently been built in city-centre locations to take advantage of transport facilities. A large stadium is used less frequently than other entertainment facilities but has a huge influx of visitors when an event takes place. Most of the positive effects of this are likely to be economic, with transport and food being the main requirements of visitors. Negative effects are likely to include congestion and pressure on public transport. As with city-centre leisure, there is also the possibility of anti-social behaviour.

The accessibility of Wembley Stadium

> **Revision activity**
>
> 1 Draw a sketch map to show the location of a major sporting event that you have studied.
> 2 Label your map to show its accessibility.
> 3 Label your map to show advantages and disadvantages for people living in the local area.

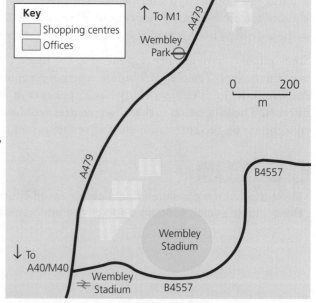

Figure 10 A map of the Wembley Stadium area in west London.

Leisure in rural areas

Most people have more leisure time than before. Transport is now also relatively cheap. This increases the **accessibility** of the countryside to urban dwellers for enjoyment. The pressures this causes to some rural areas has long been recognised, and the first UK National Park, in the Peak District, was set up in 1951. The management boards of National Parks have strong planning powers. Areas of Outstanding Natural Beauty (AONBs) are also protected but don't have their own planning boards.

> **Accessibility** The ease with which people can travel to a place

Essentially, all National Parks aim to:
- conserve and enhance the natural beauty of the area
- help the economic and social well-being of their communities
- promote the understanding and enjoyment of the National Park by visitors.

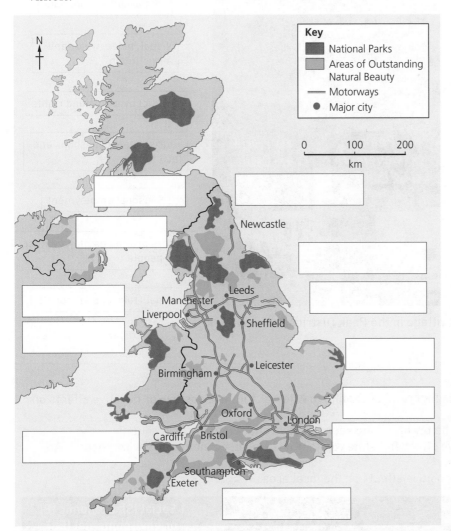

Figure 11 **The location of National Parks and Areas of Outstanding Natural Beauty.**

Revision activity

Use an atlas to help you label the map in Figure 11 to show the locations of each UK National Park.

National Park Planning Boards must manage often conflicting needs, including those of farming, providing local homes, mining, forestry, water companies, holidaymakers and conservationists. Write a list of these needs in order of most important to least important. Annotate your list to explain your two least and most important.

Exam practice

Suggest why the Peak District National Park is the most heavily used in England and Wales. Use only evidence from Figure 11.   [4] [7 lines]

ONLINE

Honeypot sites

Honeypot sites are well named. As bees are attracted to a pot of honey, people wish to visit accessible places of beauty or special interest for leisure purposes. Such visitors are not spread evenly through the year but are more frequent in the summer months and at weekends. It is easy for such attractions to reach their **carrying capacity**, beyond which they could cause a great deal of damage to landscape, ecosystems and the lives of local people. Not all effects of visitors in honeypot sites are negative, though.

> **Honeypot site** A place that attracts huge numbers of visitors
>
> **Carrying capacity** The number of visitors a place can cope with without suffering serious damage

Some effects of visitors on honeypot sites

REVISED

Footpath erosion

Road congestion

More taxes for local area

Farmers sell produce locally

Littering

Damage to crops

Increased education of visitors

Conversion of local shops to tourist trade

Disturbance of animals

Visual, noise, water and air pollution

Conflict over roaming rights

Tourists use local shops and facilities

Seasonal employment

Visitors provide employment for local people

Locals take in B&B guests

Figure 12 Castleton, a honeypot village in the Peak District National Park.

Revision activity

1 Colour code the statements in Figure 12 according to whether they are positive or negative effects on the local area.
2 Copy and complete the table below by adding each negative effect into the first column.
3 Explain why each is a negative effect and state whether it is social, economic or environmental. An example has been done for you.
4 Complete a similar table for positive effects on the local area.

Statement	Explanation	Social (S), economic (E) or environmental (En)
Littering	Causes visual pollution. Danger to wildlife and farm animals by eating it	

Managing a negative effect of honeypot tourism: footpath erosion

REVISED ☐

One of the most popular activities in National Parks and AONBs is hiking. In the summer months and at the weekends especially, large numbers of people from urban areas walk the same local footpaths.

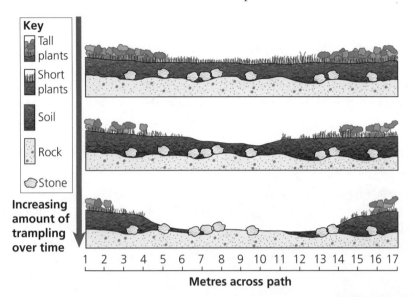

Key
- Tall plants
- Short plants
- Soil
- Rock
- Stone

Increasing amount of trampling over time

Metres across path
1 2 3 4 5 6 7 8 9 10 11 12 13 14 15 16 17

Figure 13 Causes of footpath erosion.

Tackling the problem

Apart from closing footpaths for regeneration, there are two main ways of restoring damaged footpaths:

Stone pitching: this involves digging large local stone into the ground to form solid footfalls.

Advantages	Disadvantages
Hard wearing and low maintenance	Requires skilled craftspeople
Traditional technique using natural materials	Stone is in short supply
Blends well into the surroundings	Can be uncomfortable to walk on
Suitable on steep gradients	Expensive at more than £100 per metre

Sub-soiling: this uses a digging machine to create a ditch. The sub-soil from the ditch produces a solid, hard-wearing walking surface. A specialised grass-seed mix is then sown.

Advantages	Disadvantages
Hard wearing and low maintenance	Requires experienced and skilled workers
No transport of materials is needed	Difficult access for digger
Blends very well into the surroundings	Can take several growing seasons to regrow
Comfortable to walk on	Difficult to use on paths of over 15° slope

> **Exam practice**
>
> Annotate Figure 13 to help explain changes to the footpath with time. [4]
>
> ONLINE ☐

> **Revision activity**
>
> For a location you have studied where leisure use is managed:
> a) List reasons why the location is in need of management.
> b) List ways in which the leisure use is being managed.
> c) State how effective each of these 'ways' has been.
> d) Overall, how effective has the management strategy been?

2 Urbanisation in contrasting global cities

Global cities

Global cities are the urban areas that interact the most with each other on a global scale. They are not necessarily the largest cities but are those that are best interconnected as a result of **infrastructure** improvements. All are either high-income countries (HICs) or newly industrialised countries (NICs).

> **Infrastructure** The basic services needed by a society such as water supply, sewage disposal, transport and other communications

The top 25 global cities

REVISED

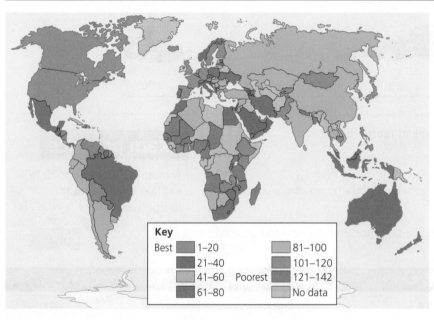

Figure 1 Global infrastructure rankings by country, 2010. A well-developed infrastructure will have strong internal transport, telephone and postal networks, and strong communication links with the rest of the world.

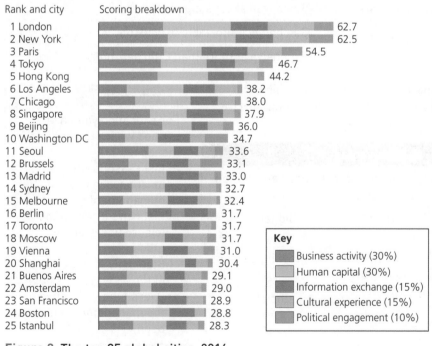

Figure 2 The top 25 global cities, 2016.

> **Revision activity**
>
> With the help of an atlas, mark and name on the map in Figure 1 each of the top 25 global cities.

> **Exam practice**
>
> a) Write three sentences that describe the distribution of the top global cities.
>   [3] [3 lines]
> b) Explain their distribution.
>   [4] [6 lines]
>
> ONLINE

Breaking down the rankings

Different groups use different criteria when determining global cities. The scores given on the previous page are based on how successfully they meet the following criteria:

Criteria	My city:
Business activity: Headquarters of global services firms Capital markets value Number of international conferences Value of goods through ports and airports	
Human capital: Size of its foreign-born population Quality of universities Number of international schools, international student population Number of residents with college degrees	
Information exchange: Access to major TV news channels Internet presence (number of search hits) Number of international news bureaux Lack of censorship Broadband subscriber rate	
Cultural experience: Number of sporting events Number of museums and performing arts venues Number of restaurants Number of international visitors Other city relationships	
Political engagement: Number of embassies and consulates Number of think tanks Number of international organisations Number of political conferences	

Revision activity

Complete this table with:
a) the name of your chosen city
b) features of this city for the five criteria above.

What does the future hold?

As **globalisation** continues, so the number of global cities will increase. Some will develop more rapidly than others but the overall pattern should be one of fewer places remaining isolated. The order of the top-ranked global cities is also changing. For example, rank 1, London, in 2016 was ranked number 2 in 2015. The internet will help you keep up with these changes. If you search for 'global cities' on the **www.atkearney.com** website, you will find the latest rankings.

Globalisation The process by which places become more worldwide connected economically, socially, politically or culturally

Two global cities ... and their challenges

You have studied two global cities:

- one in a high-income country, for example Sydney
- one in a poorer country, classified as a low-income country (LIC) or an NIC, for example Mumbai.

Use these two pages to help you write revision notes for the global city in the *poorer* country. A completed example for Mumbai is available at **www.hoddereducation.co.uk/myrevisionnotes**.

The growth and character of a global city in an LIC or an NIC

REVISED

Revision activity

On this page you need to organise the basic facts about the city by drawing the location map, explaining why it grew, and drawing two maps to show:

- the main housing zones according to wealth
- where people of different cultures live.

Annotate your maps to show the main reasons for the patterns you have shown.

City:	Country:
Location map (how it relates to its region and country)	**Why it grew** (natural population change/migration)
Socio-economic patterns and reasons: annotated map	**Cultural patterns and reasons:** annotated map

Revision activity

On this page you are going to concentrate on challenges that the local authorities are faced with and the strategies that are being used to address them.

Challenge	Strategy
Poverty and deprivation	**Self-help schemes**
Housing provision and quality	**Slum clearance programmes/housing projects**
Transport provision	**Mass transit schemes**
Growth of waste	**Waste disposal schemes**

Revision activity

You are now going to do the same for your global city in an HIC. Make a copy of page 34 and complete it for this second city. Then make a copy of the table on page 24 and in the 'Nottingham' column add details of how your chosen HIC global city is attempting to create a sustainable urban environment.

3 A global perspective on development issues

How might development be measured?

The wealth of a country is often used to measure development. It is usually expressed as its **gross national income (GNI) per capita**. This can be a useful figure but it ignores the fact that wealth and opportunity are not shared equally among all of the people in a country. Most people prefer to use the **human development index (HDI)**, because it combines a measure of wealth with social development measures.

Gross national income (GNI) per capita The total income of a country divided by its population

Human development index (HDI) A United Nations figure obtained by combining GNI per capita and measures of health, education and equality

How is the world developing?

REVISED

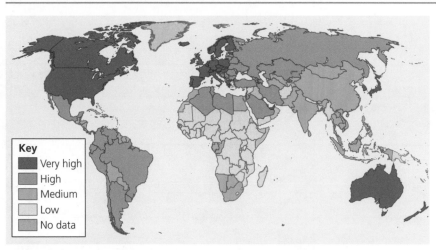

Figure 1 World HDI in 2010.

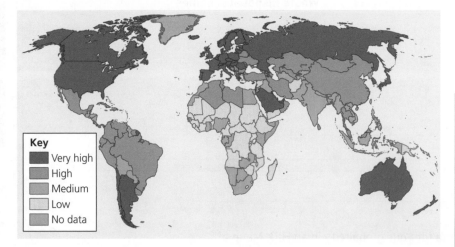

Figure 2 World HDI in 2016.

Revision activity

1 Use an atlas to help you make a list of those countries that have improved their human development status between 2010 and 2016.
2 Make another list of those in which human development has declined.
3 Why might data be unavailable for some countries?

What does this mean for people?

Healthcare

Figure 3 A nurse teaching women in a Sierra Leone village about healthcare following an outbreak of the Ebola virus. The nurse is talking to the villagers without using written information. Few if any of the women will be able to read and write. Life expectancy, 2015: UK, 81.2 years; Sierra Leone, 50.1 years.

Education

Figure 4 A US military engineer, donates books to Nigerian school children during Africa Partnership Station (APS) 2012 at the Tomaro Junior Secondary School in Lagos, Nigeria. Literacy rates (those aged over fifteen years who can read and write): Nigeria, 59.6 per cent; Russia, 99.7 per cent; UK is estimated at 99 per cent.

Equality

Saudi Arabia has strict laws about how women can behave in public. The first elections where women could vote were held in 2015 and eighteen women were elected to council seats. Percentage of female MPs at last election before April 2017: Algeria, 31.6 per cent; Rwanda, 61.3 per cent; UK, 30.0 per cent.

Figure 5 Women at a segregated McDonald's restaurant in Saudia Arabia.

Revision activity

1 Annotate the photographs on this page to show what they tell you about variations in human development. Use two or three annotations for each photo.

2 Images and quoted statistics can be found everywhere. They are mainly snapshots of a moment in time and are sometimes selected to promote a particular view to the reader or viewer. What images and information would you use if you were trying to put over your own views on human development?

Living in a global world

The key drivers of globalisation

REVISED

The good?

	Feature		Why is globalisation good?
A	It increases competition between manufacturers	1	so they are less likely to make war
B	Countries rely on each other for trade	2	so people can communicate more easily
C	Goods are manufactured in more countries	3	so aid may be organised quickly and lives saved
D	The world is becoming more and more connected	4	so higher quality products are more affordable
E	Effects of disasters are broadcast as they happen	5	so global wealth is spread more evenly

Figure 6 **Greater personal opportunity? Mo Farah, Somalian by birth.**

Figure 7 **Greater community risk? Falling water levels in wells near a bottling plant in Kerala, India.**

The bad?

- Profits made in NICs/LICs often go to HIC companies
- LICs mainly provide cheap raw materials and labour
- Local companies can't compete with incoming MNCs
- Some non-renewable resources are being used up quickly
- Western culture is spread around the world
- MNCs exploit weak anti-pollution laws in some NICs

Now test yourself

TESTED

1 Link each feature in the table with its correct 'Why is globalisation good?' explanation.
2 Add another feature and its explanation.
3 For each speech bubble, add an explanation as to why it may be a bad effect of globalisation.

Now test yourself and exam practice answers at **www.hoddereducation.co.uk/myrevisionnotes**

Is there a balance of good and bad?

Key drivers	The good	Explanation	The bad	Explanation
Increased inter-country trade				
Changes in **technology**	Encourage sharing of technology developed in high-income countries (HICs) with newly industrialised countries (NICs)	Simple innovations like wind-up radios may be used to help remote areas keep in touch with wider events	Companies who own the technology often demand high prices for their use	Governments in NICs often cannot afford the technology and lose valuable foreign exchange importing it
Changing **geo-political links**				
Cultural exchange	Exposure to the cultures of different countries leads to greater understanding between nations	This can result in an increase in respect for other people and reduces the chances of conflict		
Involvement of multi-national companies (MNCs)				

Revision activity

1 Transfer each good and bad point from page 38 to its correct column and row in the table above.
2 Add information that you have learned in your studies to your table.
3 Overall, on balance, what is your view about globalisation?

Trade The buying and selling of goods and services

Technology Means of communication, transport and manufacturing

Geo-political links Relationships between countries

Cultural exchange The movement of people and their traditions between countries

Multi-national companies (MNCs) Large companies that have their headquarters in one country and also operate in several others

An uneven globalised world

Just how even is our interconnected world? Look at the following map. In the past there would have been a dividing line on it separating the 'rich north' from the 'poor south'. One purpose of the line was to help the richer nations decide where to target **aid**. Most of the richer nations are now members of the Organisation for Economic Co-operation and Development (OECD).

> **Aid** Help given by richer societies to poorer ones
>
> **Poverty** When a person's resources are not enough to meet their needs

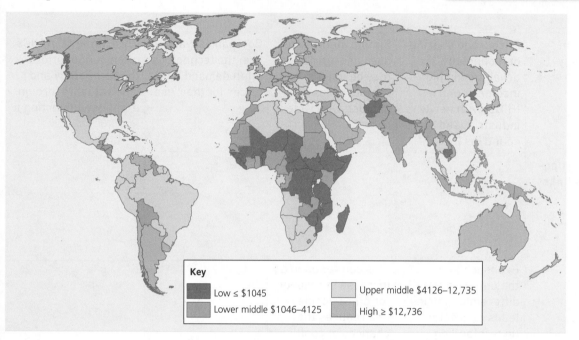

Key

■	Low ≤ $1045	■	Upper middle $4126–12,735
■	Lower middle $1046–4125	■	High ≥ $12,736

Figure 8 **The world divided by income levels.**

Regional variations

Look back at Figure 4 on page 20. As you can see, using national figures can often hide differences between different parts of a country. In this case, London's importance as a major global city has an effect on house price changes and many other aspects of life in the UK. The regional variations in UK house prices would not have been seen if we had only been given information for the country as a whole. Similarly, while a country may have a high average wealth it is likely that there will be regional variations in the percentages of people living in **poverty**.

> **The work of the OECD**
>
> We work with governments to understand what drives economic, social and environmental change. We set international standards on a wide range of things, from agriculture and tax to the safety of chemicals.
>
> We also look at issues that directly affect everyone's daily life, like how much people pay in taxes and social security, and how much leisure time they can take. We compare how different countries' school systems are readying their young people for modern life, and how different countries' pension systems will look after their citizens in old age.

> **Revision activity**
>
> 1 On the map, outline and name a low-, high- and middle-income country that you have studied.
> 2 For each of these countries, make a short list of ways in which its income level affects the lives of its people.
> 3 Suggest how the work of the OECD might be useful to people living in your low-income country.
> 4 How might the influence of the OECD be limited?

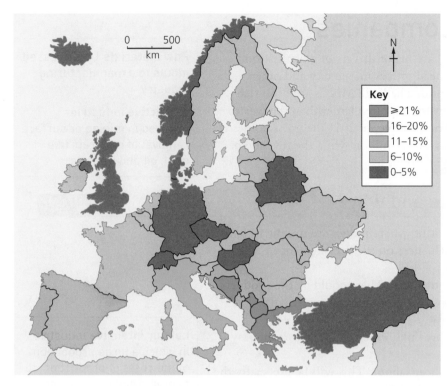

Key
- ≥21%
- 16–20%
- 11–15%
- 6–10%
- 0–5%

Now test yourself

Describe the pattern of unemployment in Europe shown by the map.

TESTED ☐

Exam practice

a) Describe the pattern of poverty in rural areas of India. [2] [4 lines]

b) How similar is the pattern of poverty in urban areas? [3] [5 lines]

c) Explain the differences you have described in question 2. [3] [5 lines]

ONLINE ☐

3 A global perspective on development issues

Figure 9 Unemployment in Europe, 2014.

It's not always just a regional thing!

REVISED ☐

We know already that much 'in-nation' migration involves movement from rural to urban areas. The perceived advantages of the cities attract large numbers from their village lives. Look at the maps below. India is one of the world's major NICs.

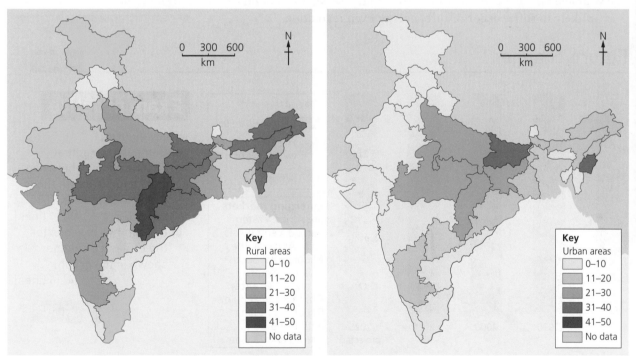

Key
Rural areas
- 0–10
- 11–20
- 21–30
- 31–40
- 41–50
- No data

Key
Urban areas
- 0–10
- 11–20
- 21–30
- 31–40
- 41–50
- No data

Figure 10 India: percentage of people living in poverty, 2012.

Multi-national companies

Multi-national companies (MNCs) are major drivers of globalisation. These large companies have their head offices in one city and other offices and factories around the world. The large MNCs are both rich and very powerful. In fact, their profits are often larger than the wealth of the world's poorest countries. For example, in 2011 the gross domestic product (GDP) of Sierra Leone was $4.215 billion while the revenue of the WalMart retail group was $485.65 billion.

> **Raw materials** Unprocessed inputs to a manufacturing industry
>
> **Extractive industries** Involved in mining or surface removal of minerals like gold, oil and iron ore

Where do MNCs locate and why?   REVISED

The reasons for any one MNC locating where it does are unique to itself. There are, though, some major influences on the decisions they make. They may be attracted by:

- A source of rare or cheap **raw materials**. These could be the products of **extractive industries** or crops that grow only in a particular climate.
- An area that has existing transport links, especially to a port for easy export.
- An area with a ready supply of cheap labour. This will be especially the case for industries that are **labour intensive**.
- Access to large **emerging markets**. NICs, in particular, are countries where the disposable incomes of much of the population are increasing, allowing more non-essential goods to be bought.
- The chance to manufacture within a large, **duty**-free, trading area like the North American Free Trade Agreement (NAFTA) or the European Union (EU).
- A friendly and stable national government, one which will give favourable tax rates and other incentives to the MNC in a country that is unlikely to suffer major disturbances or even revolution.

> **Labour intensive** Industries in which a large proportion of the costs of production go to employing the workforce
>
> **Emerging markets** Areas with large populations where there has been a great increase in earnings
>
> **Duty** A tax that is paid by a company wishing to import its goods to a country or trading zone

Future changes?   REVISED

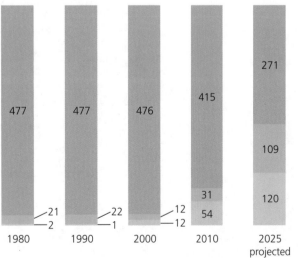

Key
- Developed regions
- Emerging markets excluding China
- Greater China

Emerging markets are Africa, eastern Europe, central Asia, Latin America, the Middle East, South Asia and South-East Asia. **Greater China** is the People's Republic of China with Hong Kong, Macau and Taiwan

Exam practice

Study Figure 11.
a) Compare the actual changes shown between 1980 and 2010 with those predicted between 2010 and 2025.   [3] [4 lines]
b) Explain two reasons for the projected changes shown in the graph.
        [4] [6 lines]

ONLINE

Figure 11 The richest 500 companies compiled by *Fortune* magazine grouped by the locations of their head offices.

Advantages brought by MNCs

MNCs will bring direct employment to an area. For example, an assembly factory will employ people in the factory and often others in the surrounding area making components. They will also create indirect employment. The benefits do not end there. Investment by an MNC is likely to bring a positive multiplier effect to the area and sometimes to the country as a whole.

Revision activity

For one MNC which is located in the UK and another located in a NIC or LIC:
● name the company
● describe and explain its choice of location(s).

Figure 12 The multiplier effect: how investment brings growth that builds on itself.

Other new businesses may be attracted to locate here

An MNC opens a new branch

Jobs are created directly within the branch

Local firms that supply the MNC with parts or services such as cleaning, maintenance or catering have more work. They take on extra staff

Local families have larger incomes. They have more spare cash

Workers can afford to take out larger loans so they might buy new cars or extend their homes

More money is spent in local shops, pubs and restaurants

Local shops, pubs and so on have more business. They may have to take on extra staff

As businesses expand they pay more tax in the form of business rates and income

Local government has more money to spend on improving roads, reclaiming derelict land and marketing the region

The image of the region is improved

The disadvantages of MNCs

The long-term records of some MNCs are not always thought to be of benefit to their host countries.

Figure 13 Photo showing protestors against Nike in the USA.

Revision activity

1 Use Figure 12 to help you draw a similar flow diagram to show what is likely to happen to an area if a large MNC factory were to close.
2 Gather evidence from these and the next two pages to help you answer the question, 'On balance, is MNC involvement good for an NIC?' Do this as a table with 'social', 'economic' and 'environmental' as headings for three rows and 'benefits' and 'disadvantages' as the columns. An example has been started below.

	Benefits	Disadvantages
Social		

Newly industrialised countries

Newly industrialised countries (NICs) are those countries which fill the gap between developed and developing countries. Imagine development as a sliding scale. Developing countries (LICs) will be at the lower end of that scale while the more developed countries (HICs) are at the upper end. At different times, since about 1950, a number of developing countries have moved out of the lower end of the scale towards the upper end. These are the NICs. Some of the earlier NICs have developed so much that they are now classified as developed countries. These include Singapore, South Korea and Taiwan. NICs develop as a result of globalisation and, especially, the involvement of MNCs. The advantages and disadvantages brought to them are, therefore, closely linked to MNC involvement.

Revision activity

1 Complete the table below for the NIC that you have studied.
2 Colour code each characteristic according to whether you consider it to be a positive or negative consequence of this development.
3 Overall, do you consider development in an NIC to be an overall advantage or disadvantage to people living in the country?

Characteristics of NICs

REVISED

NICs share a number of features which may help to explain their development or may be a result of it.

Characteristics	My NIC:
Their economies grow quickly. This is a result of a rapid increase in exports, mainly to developed countries	
Main urban areas quickly get bigger, as people migrate from rural areas seeking new jobs in the cities	
Some NICs see an increase in personal liberties and the civil rights of ordinary people. Governments control their populations less tightly	
A shift away from being a mainly agricultural nation towards more people being employed in manufacturing and construction	
A great increase in the amount of foreign investment, especially from MNCs based in developed countries	
They continue to have a cheap labour force. Sometimes this is maintained by the government banning the formation of trade unions	
They make more of the products they used to import. This cuts down the amount of money leaving the country to buy imports	
Work in manufacturing is often poorly controlled. Hours can be long and the conditions are sometimes dangerous	
More revenue from taxes allows improvements to infrastructure such as roads and airports	
Modern production methods are introduced to the country. This sometimes helps local companies to increase in size and compete internationally. Sometimes new methods are kept secret	
Participation in regional trading groups allows easier buying and selling of goods from country to country	
A rise in basic living with a resulting increase in quality of life for more of the population; usually those in urban areas	
Much of the manufacturing profit leaves the NICs and goes back to the developed countries whose MNCs control the industry	
Some MNCs have a very poor environmental record. Cases of air and water pollution are quite common	

NICs compared

The effects of globalisation and the development of NICs can be looked at in different ways. The change in overall wealth of a country (GNI) is one such way. However, this doesn't take into account how far the advantages are spread throughout the country. This can be done through examining how its HDI has changed. This measure combines wealth with life expectancy and educational opportunities.

Country	GNI per capita		HDI	
	1990	2015	1990	2015
Bangladesh (Ba)	850	3,560	0.386	0.579
Brazil (Br)	6,460	15,140	0.611	0.754
China (C)	990	14,320	0.499	0.738
India (I)	1,130	6,030	0.428	0.624
Indonesia (In)	2,840	10,690	0.528	0.689
Mexico (Me)	5,820	16,860	0.648	0.762
Philippines (P)	2,550	8,940	0.586	0.682
Thailand (Th)	9,240	15,520	0.574	0.740
Turkey (Tu)	6,030	19,740	0.576	0.767
Vietnam (V)	910	5,720	0.477	0.683

Key   High   Medium   Low

Figure 14 GNI per capita and HDI for selected NICs, 1990 and 2015.

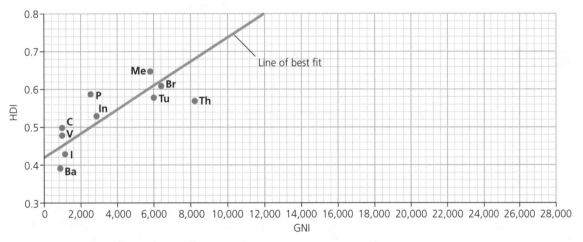

Figure 15 GNI per capita and HDI for selected NICs, 1990.

Now test yourself

Look at Figure 15.
1 To what extent does the HDI improve as the GNI increases?
2 Complete the scattergraph for 2015 using information from Figure 14.
3 'Increases in GNI have had an equal effect on all of the NICs.' To what extent does graph evidence support this statement?

Globalisation and changing world links

Globalisation will affect people differently according to where they live. These effects could be either positive or negative. Here are some possible views about how globalisation is affecting the UK.

International patterns of migration

There has been an increase in the number of workers born abroad. Migrants fill jobs ranging from surgeons to farm workers. It is argued that, with an ageing population, the UK needs young workers to help pay taxes to pay for increased pensions and medical costs. Others feel that the migrants take jobs and put increased pressure on services.

Globalisation of consumer products

The import of consumer products enables people in the UK to eat out-of-season foods all year round. It allows the purchase of, for example, electrical goods at low prices. However, some people argue that cheap imports have caused a reduction in the numbers of manufacturing jobs available in the UK and has caused redundancy.

Changes to local and national identity

Local and national identities are based on the UK's history and long experience of trade and migration with other countries. Many British people identify with the diversity created by the UK's multi-cultural population. Other people complain that the things that make the UK special or different from other places are weakened by globalisation. They want to stop future immigration, prevent cheap imports, reduce the control of the EU and create extra power for the UK government. They think that this will protect the national identity.

Globalisation of culture

People in the UK have access to TV programmes, music and computer games from around the world. We also export similar products made in the UK to many different countries either directly or in an adapted form. This brings in revenue. On the other hand, some people complain that the English language is becoming Americanised and that certain food outlets and soft-drink companies are invading our culture.

Revision activity

1 Colour code the effects of globalisation according to whether you consider them positive or negative.
2 Annotate each with S (social), E (economic) and En (environmental).
3 Add other effects on the UK for each influence.
4 Complete similar tables for your local area and for both the LIC and the NIC that you have studied.

Wider world trade issues

REVISED

Much of the world's globalisation is the result of trade. There are two basic types of trade agreements:
● free trade: the import and export of goods and services without any barriers
● restricted trade: protection of a country's or group of countries' industries by blocking products from other countries by the use of **quotas**, **import duties** or **subsidies**.

Quota A limit on the quantity of goods allowed into a country from another country

Import duty A sum of money paid when an item is imported from one country to another. Also called a tariff

Subsidy Money paid to a country's own industry to help it compete with other goods coming from outside that country

Free trade: the law of supply and demand

The **supply** of some products is dependent on factors beyond the control of producers. For example, Ghana relies on cocoa beans for nineteen per cent of its exports. The supply of cocoa depends on factors like weather conditions. Similarly, **demand** is not guaranteed to be the same from year to year. Health considerations could change the demand for cocoa. Relying heavily on primary products can cause problems for an LIC like Ghana.

Supply The amount of a product that is available for sale

Demand The amount of a product that is required by its buyers

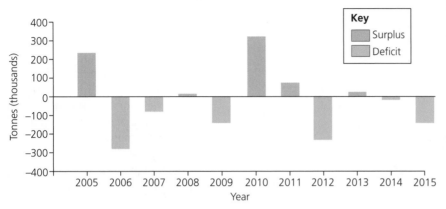

Figure 16 Cocoa supply versus demand, 2005–15.

Now test yourself

1 Label Figure 16 to show years of high and low income for cocoa farmers.
2 Suggest how the pattern is likely to affect quality of life for cocoa farmers.

TESTED

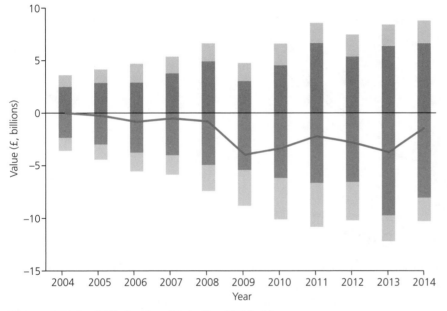

Figure 17 The UK's trade with India, 2004–14.

In 2017, the UK's trade is tied to its membership of the European Union (EU). This trade partnership uses subsidies, quotas and import duties to protect industry in its member states. This has sometimes made it difficult for countries that traditionally traded with the UK. As they develop, though, they produce more manufactured goods that are in demand in the EU. They can often still compete because of low labour costs.

Exam practice

Describe the changed trade between the UK and India shown on the graph in Figure 17.   [3] [5 lines]

ONLINE

Fair trade: a more equal world?

Fair trade organisations work to improve the lives of people, especially those in LICs, who produce primary goods for export. They attempt to get a better price for the product than would come from free trade. They also try to help farming communities to develop sustainable improvements to their quality of life. Through this they attempt to reduce global inequalities.

Figure 18 **Features of fair trade.**

> **Revision activity**
>
> For a country that you have studied, describe how the fair trade activities shown on the spokes of the wheel may help bring sustainable improvements to people living in either a cocoa-farming village in Ghana or another fair trade example that you have studied.

What is the role of aid?

When communities cannot cope without outside help, both national and international communities give assistance. This help is often given as short-term aid to solve an immediate problem caused by a natural disaster or a civil war. The aid concentrates on providing basic necessities such as food, shelter and medical help. Such help can come in the form of **bilateral aid**, **multilateral aid** and **non-government aid**. There are obvious advantages to the receiving country but the donor country often gains as well. For example, bilateral aid is sometimes linked to favourable trading deals for the donor country or political help like allowing the country's armed forces to be based in the receiving country.

> **Bilateral aid** Help from the government of one country to the government of another country
>
> **Multilateral aid** Given by governments to large international NGOs who then decide how the aid should be distributed
>
> **Non-government aid** Given by independent organisations, often charities, which collect donations to use to help countries and smaller groups

Emergency aid

The situation

More than 22 million people are affected by the drought in East Africa, and at least 13 million are experiencing severe food insecurity. The drought has caused crops to fail and cattle to die while the lack of clean water increases the threat of cholera and other diseases. Brutal war in South Sudan has driven more than three million people from their homes and left millions more in need of emergency food. There have been over 43,000 cases of acute watery diarrhoea in Somalia, Ethiopia, and Kenya.

For the first time since 2011, famine has been declared in the worst affected areas.

In Ethiopia this year, an estimated 300,000 children will become severely acutely malnourished and at least nine million people are expected not have a regular supply of safe drinking water. Over 7 million people are severely food insecure in the country.

In Kenya, over two million people are considered severely at risk. There have been some rains but the food security situation remains unchanged in the short-term.

In Somalia, more than six million people are in need of humanitarian assistance and parts of the country are on the brink of famine. Many people are relying on limited water sources which are unprotected and unsafe, exposing them to disease. The water shortage has led to reported increases in Acute Watery Diarrhoea (AWD) and cholera.

In South Sudan, over five million people - 40% of the population - are facing extreme hunger, and over 60% of the population is in need of humanitarian assistance; a humanitarian crisis driven by three years of a brutal civil war.

Without immediate action, this crisis will get worse. We need to act now to save lives.

Figure 19 Oxfam aid to East Africa, 2017.

Revision activity

Use evidence from Figure 19 and an emergency aid situation you have studied to give details of:
a) causes of situations requiring emergency aid
b) the effects of the situation on people requiring the aid
c) the response of different agencies in terms of bilateral, multilateral and non-government aid.

Development aid: which route to take to sustainability?

Development aid comes in the form of creating situations which attempt to allow communities to be able to support themselves in the future. But what form should it take? There is no simple answer to this question. In some LICs, aid has been invested in a few high-technology development projects like importing high-yield seeds, artificial fertilisers and hi-tech machinery to grow cash crops for sale abroad. In other places, small-scale intermediate technology projects have supported farming communities by ensuring a reliable water supply and providing education and healthcare in the villages.

The two examples on the next page are of aid from the Canadian International Development Agency (CIDA) to countries in Africa. Both projects were started in the early 1970s.

Mali

Aims:

- to improve village access to basic services
- to provide high-quality basic education, especially for girls
- to improve healthcare provision in the villages
- to provide more work opportunities
- to enable the provision of fair loans to farmers.

Some effects:

- Faso Jigi: an organisation with 5000 members in 134 cooperatives. It provides loans and guarantees a fair price and stable income to farmers.
- Sebenikoro Community School: 512 pupils aged from five to twelve years, of whom 320 are girls. All the teachers are male.
- National Health Sciences Training Institute: improves the effectiveness of nurses, paramedics and so on.

Tanzania

Aims:

- to set up large-scale wheat growing on traditional grazing land
- to export 50 per cent of the wheat and use the rest in a Canadian-operated bakery in Tanzania
- to use machinery imported from Canada
- to provide aid to Tanzania in return for trade with Canada.

Some effects:

- Displacement of the Barabaig pastoral nomad tribe from their traditional lands.
- Provides a source of bread consumed mainly by the urban rich.
- Biosciences Eastern and Central Africa (BECA): helps poor farmers improve their use of farming technology.

Exam practice

In the problem-solving paper, Unit 2 of your examination, you will be asked to assess the views of different groups of people and then give your own, justified, opinions. Use this opportunity to practise. Read the information above about the aid provided by CIDA as well as the quotations below:

- '… if you ask a Malian farmer what he needs, he will tell you he needs a plough, a pair of oxen and water to irrigate his fields. He will not tell you that he needs genetically modified seed.' (Ibrahim Coulibaly, Malian farmer.)
- 'Biotechnology will be the key to providing more food and other agricultural commodities from less land and water in the twenty-first century.' (Mokombo Swaminathan, Indian geneticist.)

With which view do you agree? Use detailed information from pages 48–50 and your own knowledge to help you explain why. [12] [2 pages]

ONLINE

1 Coasts and coastal management

Coastal processes and UK coastal landscapes

Being a group of islands, the UK has a very long coastline compared to its area. Many people are affected by the processes of **erosion** and weathering to which our coasts are subjected. How great the effects are depends on a number of factors that are described below.

Differing resistances of coasts to erosion

The coasts of the UK are composed of three different main groups of rocks. **Igneous rocks** and **metamorphic rocks** usually erode very slowly, while **sedimentary rocks** erode more quickly. Although not always true, the younger sedimentary rocks tend to be the least resistant to erosion. A series of glacial periods over the past half a million years have left a covering of **till** over the solid rocks of much of our coastline. These different rocks all contribute to the patterns shown on the maps below.

> ### Exam practice
>
> Explain the pattern shown by Figure 1. Use only evidence from this page.   [6] [12 lines]
>
> ONLINE

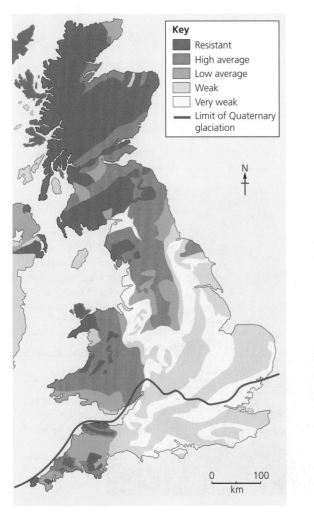

Key
- Resistant
- High average
- Low average
- Weak
- Very weak
- — Limit of Quaternary glaciation

N

0   100
km

Figure 1 Resistance of rocks to erosion in the UK.

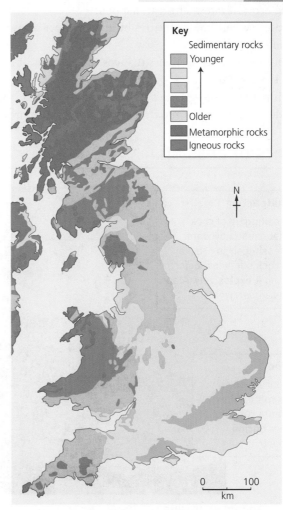

Key
Sedimentary rocks
- Younger
↑
- Older
- Metamorphic rocks
- Igneous rocks

N

0   100
km

Figure 2 Rock types of the UK.

Erosion The wearing away of rocks

Igneous rocks Formed by molten rock turning solid either on the Earth's surface or beneath it

Metamorphic rocks Formed by a change of structure of existing rocks by exposure to extremes of heat and pressure

Sedimentary rocks Formed by the erosion of existing rock and the fragments produced being cemented together to form new rock

Till A loose mixture of sediment of different sizes that is the result of erosion by ice. It is also called 'boulder clay'

Other influences on coastal erosion

REVISED

Weather

The UK normally experiences more storms and extreme weather during the winter months. These conditions of high winds and strong waves will intensify the amount of coastline destruction, especially when such weather conditions coincide with the twice-daily high tides.

Human activity

In some places, the activities of people may result in increased erosion. A build-up of sand in front of a coast helps protect it from the effects of breaking waves. Any activity that reduces the amount of sand is likely to increase erosion rates. This can happen:

- at a port, where the structures built to protect ships stop the movement of sand along a coast
- when sand mining in rivers reduces the amount of sediment being fed into the coastal areas
- when people build groynes to trap sand to protect one section of coast, it reduces the amount moving along the coast to protect that area.

Agents of erosion

REVISED

Hydraulic action
The breaking up of rock material caused by water compressing air in cracks within rock surfaces, before releasing it explosively. Also known as *quarrying*

Abrasion
The wearing away of rock surfaces by pieces of rock held in the water. Also known as *corrasion*

Attrition
The wearing down of rock material by fragments rubbing together during transport

Corrosion
The dissolving in water of soluble rock material, like limestone (calcium carbonate). Also known as *solution*

Figure 3 Agents of erosion.

How coastal landforms are produced

Differential erosion

Where rocks of different resistance are found together, the rates of erosion differ according to the resistances of the different rocks. The weaker rock will recede at a rate that is quicker than that of the stronger rock. This can be found in the headlands and bays seen in several areas around the UK coast.

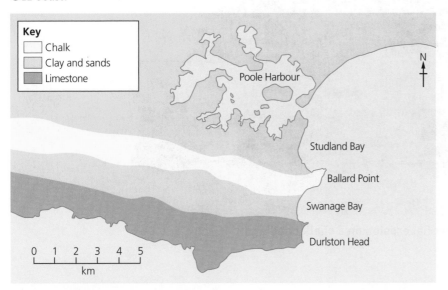

Figure 4 Headlands and bays, Dorset.

Now test yourself

1 a) Suggest an order of hardness for the three rocks shown on the map.
  b) Explain your chosen order.
2 How might geology have influenced the choice of Poole as a harbour?

Differential erosion also occurs where one rock type has zones or lines of weakness. Erosion has resulted in the following sequence taking place:

wave-cut notch → cave → arch → stack

The further the cliff retreats, the wider the **wave-cut platform** becomes.

Many stretches of coast are composed of rock that has very little resistance to erosion. The till cliffs of the Holderness coast of Yorkshire are an example. Here, the average rate of erosion is rapid, at about two metres a year. A different process assists erosion in destroying these cliffs. While waves are attacking the base of the cliff, rainfall soaks into the till from above. This bonds with the clay and increases its weight and ability to slide. Large amounts of till slide down the cliff under the influence of gravity. This process is called mass movement.

Wave-cut notch A slot cut by wave action at the bottom of a cliff

Cave A wave-cut inlet of a cliff

Arch An offshore pillar of rock linked to the mainland at the top

Stack An isolated offshore pillar of rock

Wave-cut platform A flat area of land at sea level left as a cliff retreats

Cliff attack in action

Figure 5 Differential erosion on a chalk headland.

Figure 6 Mass movement and gullying of a till cliff.

Now test yourself

1 Label Figure 5 to show three vertical lines of weakness in the photo.
2 Annotate Figure 5 to explain the sequence of erosion on this cliff.
3 Describe how geology and coastal processes have contributed to the features shown in Figure 6.

Gullying Vertical cutting in cliffs composed of weak rock and caused by rainwater erosion

Coastal transport and deposition

Transport and deposition

REVISED

When the sea or a river has eroded and broken up rock material, it then moves the rock fragments (known as the **load** when they are carried by water) to another place. It does this in four ways: in **solution** and by **suspension**, **saltation** and **traction**. If the speed of water flow is too slow to move the rock fragments, they are dumped. These processes are known as transport and deposition.

> **Load** The material carried by water as it moves
>
> **Solution** Rock material dissolves in the water. It will return to its solid state if the water evaporates

Coastal transport and deposition

REVISED

Some material eroded at the coast is carried away from the shore and is deposited at a distance. However, a large proportion of the material is transported along the coast by a process known as longshore drift. As the material rests it produces a number of features of deposition. These include beaches and **spits**.

Longshore drift

Beach material is carried up the shore at the angle of the prevailing wave movement. This is the **swash**. When the wave breaks, the water returns to the sea at right angles to the coast, pulled by gravity. This **backwash** carries the beach material with it, only to be picked up by another breaking wave, and so the process continues. It is broken when it meets a built feature like a port or a natural feature like a river **estuary**.

> **Suspension** Material 'floating' in the water as it moves
>
> **Saltation** Material being bounced along the surface by the moving water
>
> **Traction** Material being rolled or dragged along the surface by the water
>
> **Spit** A feature formed by longshore drift at a change in coastline direction when deposition continues into the sea
>
> **Swash** The flow of water up a beach as a wave breaks on a shore
>
> **Backswash** The flow of water back into the sea after a wave has broken on a beach
>
> **Estuary** Where a river enters the sea

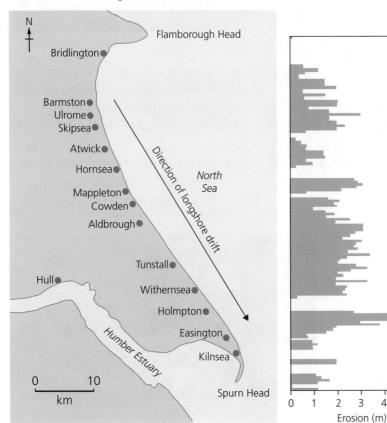

Figure 7 **The Holderness coast.**

Exam practice

'Some stretches of the Holderness coast are considered more valuable to people than others.' Support this statement using evidence from the map and graph in Figure 7.

[4] [6 lines]

ONLINE

Managing coastlines

Common coastal protection methods

There are three main ways in which people can attempt to protect the coast from wave erosion:
- by directly blocking the breaking waves by reinforcing the coastline
- by reducing the impact of the wave before it hits the coastline
- by building up the level of sediment in front of the coast to reduce wave impact.

Protection against mass movement of a cliff face requires treatment of the cliff above the zone of wave attack.

Both **hard engineering** and **soft engineering** methods are used to help defend coastlines from erosion and mass movement. All of the methods in Figure 8 below are from the Holderness coast but are also commonly used in other areas of the UK.

Hard engineering Involves building structures either parallel to the coast or at right angles to it

Soft engineering In the UK, the placing of beach material in front of an eroding coast or by regrading the cliff face

↑ **A Recurved sea wall**

1 Large material ranging from pebbles to boulders in a steel cage. Break force of waves in advance of cliffs.

↑ **B Rock armour**

2 Loose boulders placed in front of cliff or sea wall to reduce wave energy.

3 Timber or steel structures placed in front of cliff to reduce wave energy.

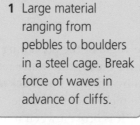

4 Inserting drainage in cliffs, reducing their angle of slope and planting grass to prevent mass movement.

↑ **C Revetments**

5 Reinforced concrete structures with an outward curved upper part to deflect wave energy back to sea.

↑ **D Gabions**

↑ **E Cliff stabilisation**

Figure 8 Methods used to help defend coastlines from erosion.

Costs and benefits of coastal protection methods

All of the methods we use to attempt to protect our coastline have both advantages and disadvantages. Each of these may be subdivided into social, economic and environmental benefits or costs. Consider the table below:

Method	H or S?	Effect		
		Social	Economic	Environmental
Sea wall, for example		Acts as a promenade Blocks views of sea May restrict access to beach	Attracts tourists Costly to build and maintain	Can cause break up of beach Deflected water can undercut base of wall
Revetment, for example				
Gabions, for example		Visually unappealing Restrict access to beach	Relatively cheap to build Limited lifespan Deter tourists	Hold beach material in their structure Ineffective in storms
Groynes, for example		Build up beaches Can look unappealing	Cost effective Little maintenance	Reduce wave effectiveness Starve downdrift beaches
Cliff stabilisation, for example			Relatively cheap Short term	Prone to cliff slumping
Beach nourishment *Importing sand, usually dredged offshore, to the beach, for example*			Attracts tourists Requires constant replenishment	A useful supplement to groynes
Beach stabilisation *Planting dead trees in the beach to help stabilise and widen it, for example*				
Wetland creation or protection *In the UK usually involves salt marshes, for example*		Recreational value	Relatively cheap Attracts visitors	Absorbs wave energy Protects valuable habitats

Coastal protection at Mappleton

REVISED

To protect a particular stretch of coast or not is a major decision for both national and local governments. They also need to decide how much money to spend on the protection. Much depends on how valuable the stretch of coast is thought to be. Mappleton (see Figure 9) is a village of about 350 people. At Mappleton, a combination of relatively cheap hard and soft measures have been used.

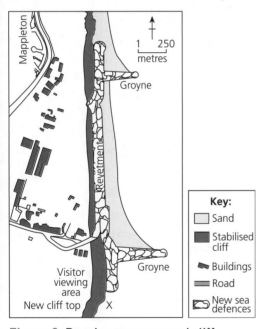

Figure 9 Beach processes and cliff protection at Mappleton.

Figure 10 Looking north from the south groyne at Mappleton.

Now test yourself

TESTED

1 Write a definition of 'groyne'.
2 Label Figure 9 to show the direction of longshore drift.
3 Annotate Figure 9 to show how hard and soft engineering is being used to protect this stretch of coast.

Exam practice

Explain why erosion has increased at point X on the map in Figure 9 since the creation of the protection scheme. [4] [6 lines]

ONLINE

Hold the line or managed retreat?

Who makes the decision?

There is no legal duty for the government to protect people and their property from coastal erosion and flooding. The national government formulates the overall strategy but each local authority in England and Wales must prepare a **Shoreline Management Plan (SMP)** for its section of coast. Local authorities are advised by the Environment Agency and local coastal erosion risk management authorities when developing the SMP.

As protecting the coast is expensive, local authorities must produce a **cost-benefit analysis** which may consider such factors as:
● the value of the property that is threatened
● how many people live along the stretch of coast
● the cost of replacing infrastructure destroyed by the eroding coast
● the effect of destruction on the overall economy of the area.

Doing nothing at all is sometimes considered, as well as the possibility of building new coastal defences out to sea. However, the choice usually comes down to either '**hold the line**' or '**retreat the line**'.

> **Shoreline Management Plan (SMP)** A detailed set of strategies for the future management of a stretch of coastline
>
> **Cost-benefit analysis** A consideration of the balance between the advantages and disadvantages of implementing a particular strategy
>
> **Hold the line** Using coastal protection methods to prevent erosion
>
> **Retreat the line** Accepting that erosion will take place and putting strategies in place that will protect people as it happens. Also called managed retreat

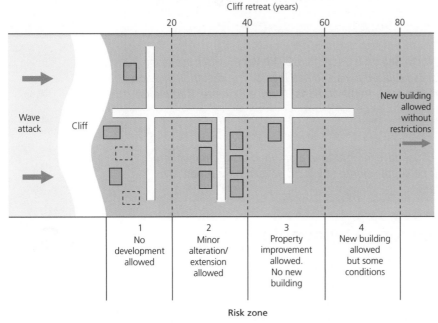

Figure 11 'Retreat the line' in action.

Should the Holderness coast be protected?

If the coast is not protected, it will continue to erode at the present rate (see Figure 7, page 55). There have been over 20 villages lost to the sea along this coast since Roman times. This trend will continue. Protecting the whole coastline is very expensive and not an option. As you have already seen, the small village of Mappleton has already been protected using relatively cheap methods (Figure 9, page 58). In contrast, the holiday resort of Withernsea has a concrete sea wall, protected by rock armour and a series of wooden groynes.

Different views

Different stakeholders in the area will have different views on the issue. These will reflect their own circumstances and the ways in which these circumstances influence the **value** positions they hold on the issue.

> **Values** The things people believe that are important to the way they live. Our values inform the attitudes we have to different issues

Stakeholder	For or against protecting the Holderness coast
Mappleton resident *Before we were protected, the local erosion rate was two metres a year. By 1988, my bedroom was eight metres from the cliff edge. Since the scheme was put in place there has been no more erosion*	
Cowden resident *We could just about cope with the erosion before the Mappleton protection scheme was put in place. Now erosion has increased greatly and there is danger to both our settlement and the caravan park between it and the sea*	
Withernsea shopkeeper *Occasionally, the sea wall is breached, but it is then repaired. Although not as popular as in the past, we still get many summer visitors and the town is also a retirement centre*	
Easington gas terminal worker *Gas from Norwegian fields in the North Sea is brought ashore here. The terminal is protected by rock armour. The cliff has been re-profiled, covered with mesh and grass covered. There is no groyne at the southern end and little evidence of increased erosion rates to the south*	
Spurn Point resident *A storm surge funnelled down the North Sea on 5 December 2013. Very low air pressure and a narrowing of the sea southwards resulted in raised sea levels. The resulting erosion washed away a large section of the road down to the National Nature Reserve. Rebuilding this road is unrealistic*	
Hull resident *£80 million has been granted by national government to use on flood protection schemes along the Humber Estuary. This will be supplemented by local government funds. Over 8000 Hull homes were flooded in 2007*	

Historic changes to coastlines by people

REVISED

People have a long history of changing coasts for economic benefit. Our coasts have many harbours where boats rest in the shelter of sea walls. Large industries have also modified the coastline to manufacture goods based on imported raw materials like iron ore and oil. These changes always have some effect on the normal processes of erosion, transport or deposition operating along that stretch of coast.

Exam practice

Study the information on pages 55–60. How should the local government respond to the issue of coastal erosion along the Holderness coast?
● Approach 1: adopt a policy of retreat the line.
● Approach 2: adopt a policy of hold the line using hard engineering methods.
● Approach 3: adopt a policy of hold the line using soft engineering methods.

Draft a letter to the East Riding of Yorkshire Council. Explain why your chosen option or options should be adopted. Justify your decision by referring to information on these pages and your own knowledge.

Your ability to spell, punctuate and use grammar and specialist terms accurately will be assessed in your answer to this question.

[12+4] [1.5 pages]

ONLINE

The water cycle: a system of stores and flows

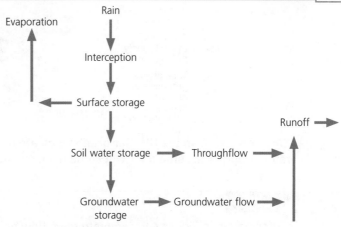

Figure 1 **The hydrological (water) cycle in action.**

> **Stores** Where water remains in one place or the same physical state (liquid, solid, gas)
>
> **Flows** The transfer of water between stores, in the same state or involving a change of state

Revision activity

1 Add the following **stores** to Figure 1: ice cap, sea, woodland, lake, groundwater.
2 Label the diagram with **flows** 1 to 6.

	Flow	Definition
1	Interception	Falling water hitting the Earth's surface
2	Evapotranspiration	Transfer of water vapour from the Earth's surface and from vegetation
3	Infiltration	Water passing into the soil from the Earth's surface
4	Surface runoff	Water flowing across the Earth's surface, usually as rivers and streams
5	Throughflow	Water moving slowly through the soil
6	Groundwater flow	Water moving through the rocks

Natural influences and human interference

REVISED

Both vegetation (natural and farmed) and geology will influence the rate at which water passes through the cycle:

- plant cover increases transpiration but usually decreases surface runoff
- steeper slopes increase amounts of surface runoff
- permeable rocks encourage more infiltration than impermeable rocks
- porous rocks act as stores of water.

Modern towns and cities create an environment in which much of the water falling on them bypasses stages involved in a natural cycle.

Figure 2 **Rural discharge.**

Figure 3 **Urban discharge.**

TESTED

Now test yourself

Look at Figure 3 on page 61. Explain why each of the following may affect an urban discharge pattern: paved gardens, interception of water by buildings, street drainage.

Providing water: a major influence on the water cycle

REVISED

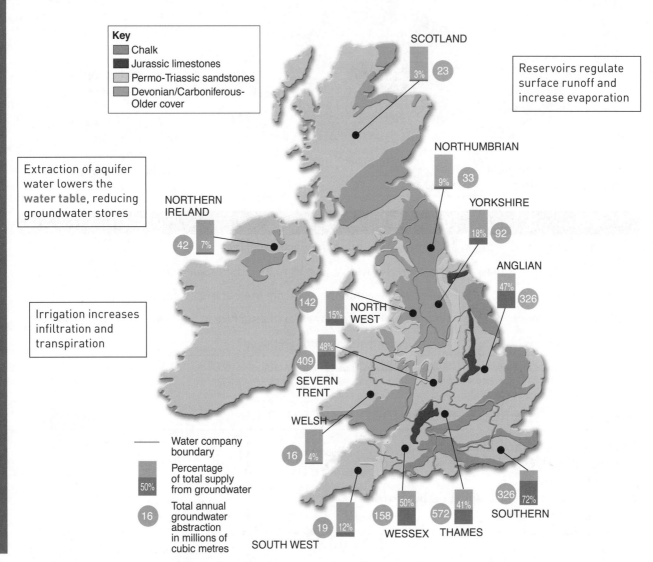

Key
- Chalk
- Jurassic limestones
- Permo-Triassic sandstones
- Devonian/Carboniferous-Older cover

SCOTLAND 3% 23

Reservoirs regulate surface runoff and increase evaporation

NORTHUMBRIAN 9% 33

Extraction of aquifer water lowers the **water table**, reducing groundwater stores

YORKSHIRE 18% 92

NORTHERN IRELAND 42 7%

ANGLIAN 47% 326

142 NORTH WEST 15%

Irrigation increases infiltration and transpiration

409 SEVERN TRENT 48%

WELSH

16 WELSH 4%

— Water company boundary

50% Percentage of total supply from groundwater

16 Total annual groundwater abstraction in millions of cubic metres

19 SOUTH WEST 12%

158 WESSEX 50%

572 THAMES 41%

326 SOUTHERN 72%

Figure 4 Major **aquifers** in the UK. Bar graphs indicate the percentage of water supply that comes from groundwater supply. The remainder will come from surface stores (rivers and reservoirs).

Revision activity

Use the information in Figure 4 to complete the following sentence: 'My water company is _____ . Its percentage of water from groundwater sources is _____. Its total annual groundwater abstraction is _____ million cubic metres. This means that my region supplies a total of _____ million cubic metres of water a year.

Water table The upper level of groundwater

Aquifer Store of water in porous rocks that is extracted for human use

The relationship between climate and river discharge

The **annual regime** can be shown on a hydrograph. Hydrographs show the relationship between precipitation and river discharge. The high and low discharge periods will correspond to periods of greater and lesser precipitation, although snow accumulation and melt may complicate this simple picture. It is important to realise that the annual regime is for just one particular year, whereas climate figures and average river flows are the result of data collection over longer periods; usually periods of at least 30 years. See also page 66 for the effects of changes in discharge rates on people over relatively short periods of time.

> **Annual regime** Pattern of water discharge throughout the year

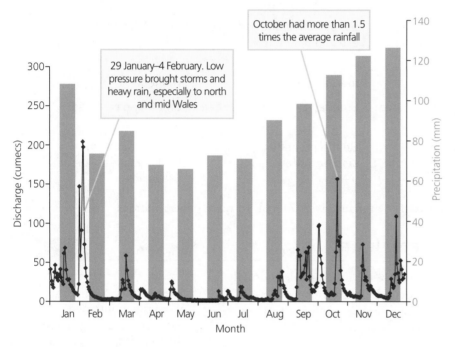

October had more than 1.5 times the average rainfall

29 January–4 February. Low pressure brought storms and heavy rain, especially to north and mid Wales

Figure 5 Hydrograph for the River Dyfi, Wales, 2004.

Revision activity

Make a list of ways in which the activities of people affect the water cycle in your area.

Exam practice

a) Compare the annual regime of the River Dyfi with the rainfall pattern. [3] [5 lines]

b) Explain why there are differences between the two patterns. [3] [5 lines]

ONLINE

River processes and UK river landforms

River erosion, transport and deposition

REVISED

The same processes of erosion operate along the course of a river as are found along the coast (see page 52). This is also true for transport and deposition (see page 55).

The dominance of erosion, transport and deposition processes change as you move down a river's **long profile** from source to mouth.

Feature	Course of river		
	Upper	**Middle**	**Lower**
Gradient	Steep	Gentle	Very gentle
Channel	Shallow and narrow	Wider and deeper	Wide and deep
Processes	Much friction and so rapid erosion. Attrition has not had much effect yet	Less friction and reduced erosion. Much transport	Little friction. Mainly transport and deposition
Discharge	Low	Increasing	Highest
Bedload	Large. Sub-angular stones and boulders	Smaller and more rounded	Small and rounded
Landforms	Interlocking spurs. V-shaped valleys. Rapids and waterfalls	Gentle meanders. Gently sloping valleys	Wide meanders. Ox-bow lakes. Floodplain. Estuaries

While the relationships in the table are true of many rivers, not all fit these generalisations.

V-shaped valleys and waterfalls

REVISED

In the upper course of a river, the gradient of the long profile is usually steep, resulting in most erosion being vertical. This produces narrow valleys with steep V-shaped sides. When a river passes over a band of more resistant rock, a waterfall is likely to form.

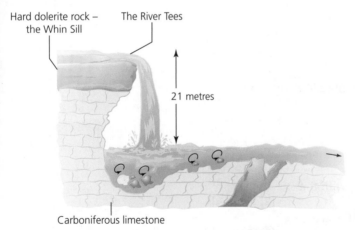

Hard dolerite rock – the Whin Sill

The River Tees

21 metres

Carboniferous limestone

> **Long profile** A line graph along the length of a river showing its gradient changes from source to mouth
>
> **Discharge** The volume of water flowing through a section of river at a given time. It is measured in cubic metres per second (cumecs)
>
> **Bedload** The material that is carried by a river by traction and saltation

Figure 6 How the waterfall and gorge at High Force are formed.

Over many years, the waterfall will erode back, cutting a narrow, steep-sided valley known as a gorge.

Revision activity

Describe how processes of erosion have combined to create the waterfall in Figure 6.
1 Draw a sketch of Figure 6.
2 Label your sketch with the following features: cap rock, plunge pool, less resistant rock.

3 Annotate your sketch to show the effects of the following processes: abrasion, hydraulic action, attrition, corrosion, gravity.
4 How might the processes operating on the waterfall result in the formation of a gorge?

Meanders and floodplains

As a river flows into its middle and lower courses, the gradient of the long profile becomes shallower and the river starts to bend and travel across a wider valley. The **river channel** widens. Much of the bedload carried from higher up the river is deposited. The following sequence takes place to form a meander and a floodplain.

1 Water flow is fastest on the outside of the bend. Erosion creates a **river cliff**.
2 Water flow is slowest on the inside of a bend. Deposition creates a **point bar**.
3 A combination of 1 and 2 causes the river to move, or meander, across its valley.
4 In times of high rainfall or snow melt, the river flows onto the valley floor and deposits material to form the floodplain.
5 Deposition is usually greatest close to the river banks. This forms a **levee**.

> **River channel** The area that contains the flowing water
>
> **River cliff** A steep-sided river bank
>
> **Point bar** A shallow river beach, also called a slip-off slope
>
> **Levee** A river bank higher than the level of the flood plain

Estuaries and ox-bow lakes

- Estuaries form where rivers enter the sea (marked E on Figure 1, page 61). Many features are similar to the lower course of a river. However, deposition of the finest material results in mud flats and salt marshes. The estuaries are affected by twice-daily tides in addition to the flow of the river. The estuary landscape is constantly changing.
- Ox-bow lakes are created when the inside bends of a meander join. The water flows straight across, cutting off the meander to leave it as an ox-bow lake.

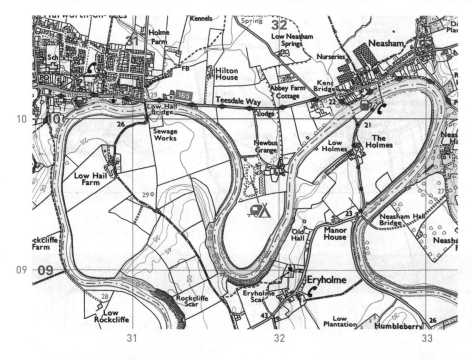

Figure 7 The River Tees, near Hurworth-on-Tees. Scale 1:25,000.

Now test yourself

Draw a sketch map to show where an ox-bow lake may form. Annotate your map to describe how processes of erosion and deposition may lead to its formation.

Causes of river floods

Quite simply, rivers flood when there is too much water flowing down them to hold. When the cumecs (cubic metres per second) are greater than at **bankfull stage** the excess water flows onto the floodplain.

Flood hydrographs

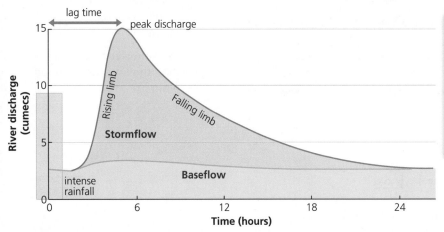

Bankfull stage When a river fills its channel

Baseflow The normal rate of discharge of the river

Stormflow The rate of river discharge following a period of abnormally high rainfall

Figure 8 A simple flood hydrograph.

Revision activity

1 Complete the table by adding definitions of each of the terms.

2 Bankfull stage for this river is a flow of 12 cumecs. Add this to Figure 8.
3 For how long will the river flood?

Term	Definition
Lag time	
Peak discharge	
Rising limb	
Falling limb	

Factors affecting flooding

In the example above, the period of flooding was caused by intense rainfall. In severe instances there can be flash floods, like those experienced in Boscastle in 2004. There are, though, a number of other natural and human reasons why a river may flood. Some of these factors are natural while others are the result of human activity.

Human or physical?	Factor	Explanation
	The soil is saturated from previous rainfall	
	Removal of forests causes rapid soil erosion	
	A very steep drainage basin	
	A wide drainage basin with many tributaries	
	Storm drains feed directly into the river system	
	Heavy rainfall in early spring	

All of the factors above will result in an increase in river discharge, sometimes reaching bankfull stage and causing flooding.

Revision activity

1 Complete the table on the previous page by stating which factors are human and which are physical.
2 For each factor add the correct explanation from the list below into the correct column in the table:
   - *so* water is quickly fed into the river
   - *so* infiltration is blocked and surface runoff increases
   - *so* large amounts of water drain to the river
   - *so* the riverbed becomes shallower
   - *so* direct rain is supplemented by snow melt
   - *so* tributary flow is rapid to the main river.

Seasonal flooding

Flash floods are the result of extreme weather events, such as an intense period of rainfall. There are some places, though, where flooding is a regular occurrence. One such place is the Mekong Delta in Cambodia. It suffers from the effects of severe storms called cyclones between mid-April and mid-October. Cyclones are large air masses with very low air pressure, which causes local sea levels to rise. During the cyclone season the delta floods. The floodwater is the result of:

River discharge from rain and snow melt (60%) + direct heavy rain (10%) + high tidal flow from the sea (30%)

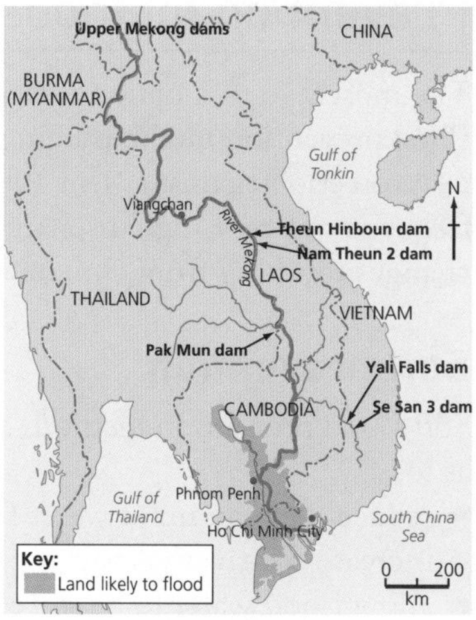

Exam practice

1 Draw a graph to show the composition of the Mekong Delta floodwater. [3]
2 Give two advantages of your choice of graph. [2] [2 lines]

ONLINE

Figure 9 Flooding in the Mekong Delta.

Living with floods

People used to flooding plan their lives around it. In the delta area, there are healthcare boats, life vest distribution, a flood warning system and evacuation plans. Trees are planted along raised roads, houses are on stilts and many crops are harvested before the floods. Rice and fish are farmed. However, flooding is occasionally severe. One such year was 2011.

Type of loss	Loss and damage in the delta area
Human	89 people died
Built features	900 houses destroyed; 176,000 houses flooded; 1200 classrooms damaged
Farming	27,000 hectares of rice padis and 74,000 hectares of fruit trees ruined
Fisheries	7300 hectares of fish farms and over 5000 fish cages damaged. 5600 tonnes of shrimp and fish lost
Transportation	554 km of national roads and 316 km of rural roads destroyed; 9000 bridges destroyed
Water control structures	4 million metres of dykes and 1.2 million metres of irrigation canals destroyed

Revision activity

1 Use information on this page to create a list of *social* positive and negative effects of flooding in the Mekong Delta.
2 For each effect write a 'so' statement to help explain why it is either negative or positive.
3 Repeat this activity for *economic* effects.

Responding to river floods

There are two ways in which people might respond to river floods:
● by protecting against their effects when they occur
● by preventing them happening.

Protection methods

The responses shown in red type on page 67 are aimed at protection against the effects of floods when they occur. It is also possible for individuals and local governments in the UK to protect both people and their property by methods like moving belongings upstairs and sandbagging the entrances to homes.

Prevention methods

The most favoured option is to stop the floods happening. Most new flood prevention methods being built are expected to be sufficient for protection against a '1 in 100-year' flood. This likelihood of the frequency of very high discharge rates does not take into account possible changes resulting from climate change.

Soft prevention

Soft flood prevention methods alter the catchment area conditions so as to:
● reduce the amount of water flowing into the river
● increase the river's capacity
● allow flood water to be stored naturally on the floodplain.

They reduce the risk of flooding along one section of the river without increasing the risk lower down its course. Some of these natural flood management methods are shown in the diagram below.

The effect of such measures on a hydrograph would be to increase the lag time, reduce the level of peak discharge and lengthen both the rising and falling limbs.

> ## Now test yourself
>
> Describe the effects of each of the methods shown in the diagram using the following terms: infiltration, evapotranspiration, surface runoff, soil erosion, interception, river channel.
>
> TESTED

Blocking upland drainage

New wetlands in the upper catchment

Floodplain water storage to reduce flow

Fence to keep livestock from river bank

Reduce amounts of bare soil

Increase woodland on the floodplain

Figure 10 Natural flood management methods.

Hard prevention

Hard flood prevention methods are aimed at stopping the water from breaking its banks or, if it does, stopping it covering areas of high value to people. Deepening the river channel increases the volume of water it can carry without flooding. Raising embankments or building walls has the same effect. Straightening the channel speeds the flow of water past the protected area. However, using these hard methods, preventing flooding along one section is likely to increase the amount of flooding lower down the river's course.

The effect of such measures on a hydrograph would be to decrease the lag time, increase the level of peak discharge and shorten both the rising and falling limbs.

Finally, operating alongside both sets of prevention measures is the idea of zoning floodplain land. Those areas of low value like sports fields, golf courses and car parks are not protected and will be allowed to flood. Planning permission will not be given to build on them.

> **Revision activity**
>
> For each method in the table, make a list of other costs and benefits you have studied.

Costs and benefits of different types of flood prevention methods

REVISED

Method: soft/hard	Effects	Costs	Benefits
Tree planting and crop growth	Increases evapotranspiration Slows surface runoff Reduces river flow	Loss of farmland and jobs	Produce sold and jobs created. Environment sometimes improved
Protection of river banks	Prevents bank erosion Reduces river transport load	Lost use of some grazing land	Increases carrying capacity of river downstream
Managed flooding	Stores excess flood water Releases it slowly Equalises river flow	Takes land out of year-round use	A natural-looking barrier
River straightening	Speeds river flow past protected area Increases erosion	Disrupts anglers and other leisure users. More silting further downstream	Sale of extracted gravel
River deepening and widening	Increases river carrying capacity near protected area	Reduces river levels in non-flood times – disrupts leisure use of river	Sale of extracted gravel. New habitats created at river's edge
Dam construction	Stores excess flood water Controls its release Equalises river flow	Loss of farmland. Housing and communities destroyed	Water for irrigation and electricity generation. Leisure use
Raising river banks	Increases river carrying capacity near protected area	Loss of river views. Disrupts river access. Increases flow to places downstream	Increases local property values. Reduces insurance costs
Flood plain zoning – a protection method	River floods naturally Different types of land use banned dependent on risk	Loss of land for housing development	Low-cost land for reserved for recreation and car parks

Key: Increasing financial cost: **green** → **orange** → **red**.

Views on drainage basin management

As with the coastal protection exercise you were asked to complete on page 60, people have different views about protection against river flooding.

What would you have done?

The village of Attenborough, a commuter village to the west of Nottingham, received a new flood wall. Not all the villagers had the same views as to where it should be built.

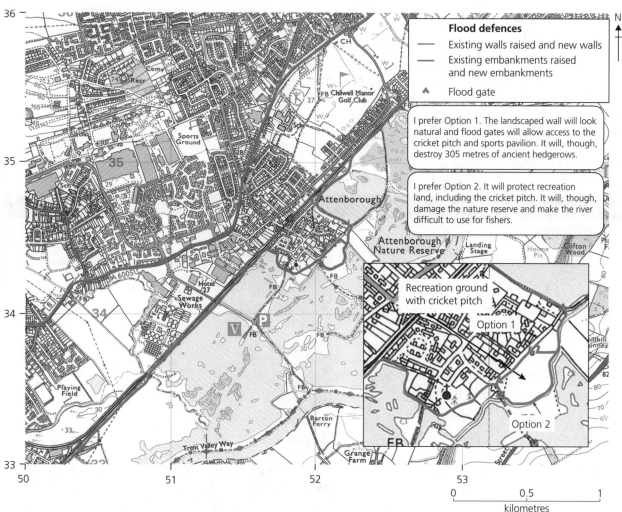

Flood defences

— Existing walls raised and new walls

— Existing embankments raised and new embankments

▲ Flood gate

I prefer Option 1. The landscaped wall will look natural and flood gates will allow access to the cricket pitch and sports pavilion. It will, though, destroy 305 metres of ancient hedgerows.

I prefer Option 2. It will protect recreation land, including the cricket pitch. It will, though, damage the nature reserve and make the river difficult to use for fishers.

Recreation ground with cricket pitch

Option 1

Option 2

Figure 11 **OS map showing the new flood wall on the River Trent. Scale 1:25,000.**

Exam practice

Which of these three options would you have chosen to protect Attenborough? Consider the social and economic costs and benefits on people in the village and on settlements further downriver. Choose from:

- option 1 on the map
- option 2 on the map
- option 3, to use 'natural' methods higher in the river catchment.

Draft a letter to Nottinghamshire County Council. Explain why your chosen solution should be a priority. Justify your decision using information from these two pages.   [12+4] [36 lines]

Your ability to spell, punctuate and use grammar and specialist terms accurately will be assessed in your answer to this question.

What actually happened?

Figure 12 **The cricket pitch before building began. November 2000: NE from GR. 520344.**

Figure 13 **The cricket pitch on completion of the flood wall. February 2013: from the same location.**

Historic changes to rivers by people

REVISED

People have a long history of changing rivers for economic benefit. During the Industrial Revolution, water power was used by early factories. Rivers had their courses changed and reservoirs were built to store water for use in manufacturing. This continued when steam power took over as the water was still needed. These changes are still in place, often unused, and continue to affect the processes of river erosion, transport and deposition. In more modern times, river water continues to be rerouted to cool electricity power stations. An additional effect of this is the heating of rivers by water returned to them. This impacts on river ecosystems.

3 Weather and climate

Weather and climate

Influences on the UK climate

REVISED

The weather we experience depends on a number of factors. These are each quite simple but when combined can lead to a very complex picture that gives each region a unique character.

Characteristic	Regional variations
Global atmospheric circulation: the weather we get depends on the position of the **polar jet stream** (page 74) and its influence on the air mass we experience at any one time. Due to the Earth's rotation, weather systems usually pass over the UK from west to east	As the jet stream is constantly changing position and the weather systems that follow it also shift, it is not possible to make simple statements about regional differences. The map in Figure 1 shows the seasonal weather we might expect from each of the air masses that influence our weather
Air pressure: in low-pressure situations rising air cools and as it does it loses its ability to retain moisture as a gas. The water vapour condenses to form water droplets. These join together and, when large enough, fall as rain or snow High pressure produces sinking air. This heats as it sinks, increasing its ability to hold water vapour. Such air is 'stable', with no precipitation. As skies are often clear, they bring cold conditions in winter and hot in summer. Diurnal temperature ranges are high	As low-pressure systems blow off the Atlantic Ocean from west to east, it is the west that experiences the greatest rainfall. This is helped by the fact that the air rises to travel over hills and mountains before sinking down their eastern sides High-pressure areas (anticyclones) often settle over Europe for weeks. Their influence, therefore, is felt a little more in the east of the UK than the west
Latitude: this is the distance a place is from the Equator. Because the Earth is curved, the angle at which the Sun's rays hit the Earth (the angle of incidence) reduces with distance from the Equator. This means that the heating effects also reduce. This loss is further affected by the greater distance the rays have to pass through the atmosphere causing increased reflection back to space	In both summer and winter, the north of the UK is colder than the south. This is partly the result of latitude but the influence of altitude is at least as important. In fact, there is also an east to west difference, with the east, on the whole, being warmer than the west. Latitude is of great importance when considering differences between areas close to and far from the Equator. The 600-mile length of the UK (just over 8° of latitude) is of less influence
Altitude: the height above sea level of a location. With height, air becomes less pressurised so incapable of holding so much heat. The loss is 9.8 °C for every 1000 m under a clear sky but only about 3.3 °C when in cloud. This is due to the ability of the increased proportion of water vapour in the atmosphere to absorb heat	Highland regions will be colder than lowlands. Loosely, this suggests that the influence of latitude will reduce the further south-east one travels. Two places geographically very close to each other can have greatly different temperatures At 16:00 on 25 February 2017 the forecast for Ben Nevis (1344 m) was −4 °C while, at its foot, Fort William (24 m) was +8 °C, a temperature range of 12 °C

Characteristic	Regional variations
Distance from the sea: sea areas are less effective conductors of heat than land areas. So they have a lower range of temperatures and winds blowing off the sea will have a modifying effect on coastal areas. In winter, they will help make such areas milder. In summer, they will have a cooling influence. These effects are lost with distance from the sea	As the UK is an island, the simple influence here is that the further inland a place is, the warmer it will be in summer and the colder in winter. However, we are also influenced by our nearness to the land mass of mainland Europe, where summer temperatures in the east are very high and winter temperatures very low. The effect of this is sometimes seen in the eastern UK being colder in winter and hotter in summer than the west

Air masses affecting the UK

The UK is affected by a number of different **air masses**. These bring quite different weather to people in the UK depending on which air mass is travelling across at any time. This is mainly controlled by the track of the polar jet stream. This is a powerful high-altitude wind found between 9 and 12 km above the Earth. It is the interface between high- and low-pressure air masses. The track of the jet stream migrates north and south throughout the year. If it flows to the north of the UK, there is a tendency for tropical air masses to affect us. However, when its track is further south, there is more likelihood of experiencing polar air masses.

Polar jet stream A band of strong winds between 9 and 12 km above the Earth. It forms along the interface between high- and low-pressure air masses

Air mass A large volume of air having the same temperature and moisture properties

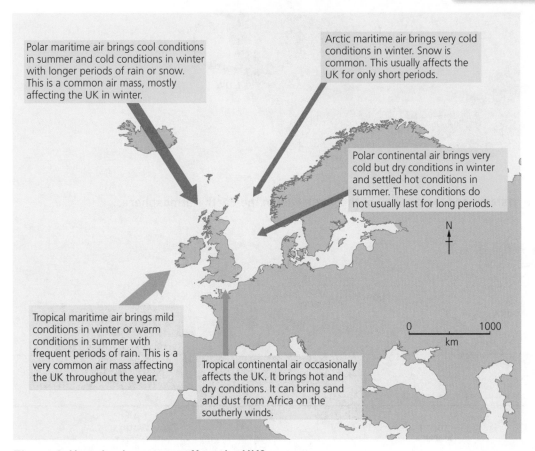

Polar maritime air brings cool conditions in summer and cold conditions in winter with longer periods of rain or snow. This is a common air mass, mostly affecting the UK in winter.

Arctic maritime air brings very cold conditions in winter. Snow is common. This usually affects the UK for only short periods.

Polar continental air brings very cold but dry conditions in winter and settled hot conditions in summer. These conditions do not usually last for long periods.

Tropical maritime air brings mild conditions in winter or warm conditions in summer with frequent periods of rain. This is a very common air mass affecting the UK throughout the year.

Tropical continental air occasionally affects the UK. It brings hot and dry conditions. It can bring sand and dust from Africa on the southerly winds.

N

0   1000
km

Figure 1 How do air masses affect the UK?

Global climate zones

Northern hemisphere controls

The different **climate** zones around the globe are the result of the differences in the heating effects of the Sun between the Equator and the Poles. A similar circulation to Figure 2 exists in the southern hemisphere. It is the activity of the different air masses that this global circulation brings that results in different climate zones.

Climate Average weather conditions taken over a period of at least 30 years

Insolation rate The amount of solar energy received per square centimetre per minute

Diurnal range The difference between night-time and daytime temperatures

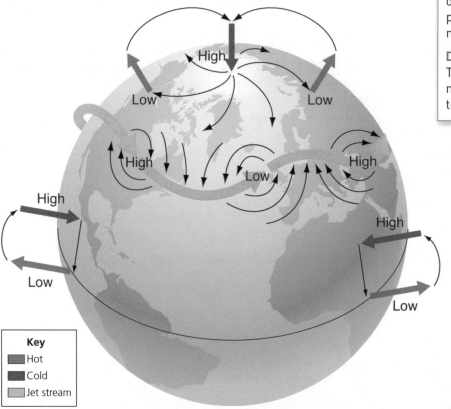

Figure 2 Pressure systems and the position of the jet stream in the Earth's atmosphere.

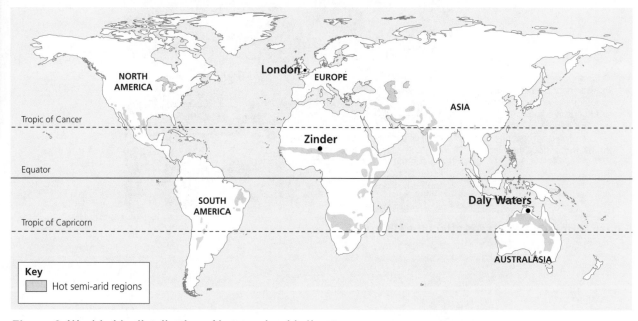

Figure 3 Worldwide distribution of hot semi-arid climates.

Two climate zones compared

Hot semi-arid	
Distribution On the edges of arid areas in bands closer to the Equator. Almost wholly between the two tropics. They are in a zone of changed atmospheric conditions between hot deserts and areas having a seasonal tropical wet and dry climate	**Distribution**
Temperature distribution	**Temperature distribution**
Temperature reasons The Sun is high in the sky during all twelve months. Skies are also cloudless. This results in a high **insolation rate** so there are hot conditions throughout the year. However, heat loss in great at night and the **diurnal range** is relatively high	**Temperature reasons**
Precipitation distribution	**Precipitation distribution**
Precipitation reasons As temperatures are high, precipitation falls only as rain, unless at very high altitude. The dry period results from high-pressure systems that have descending air that is stable. Much precipitation in the wet period is the result of thunderstorms	**Precipitation reasons**
Climate graph Key — Temperature ▨ Rainfall	**Climate graph**

Rainfall (mm) / Temperature (°C) — Month: Jan Feb Mar Apr May Jun Jul Aug Sep Oct Nov Dec

---

Revision activity

You have studied another climate as well as the hot semi-arid climate.
a) Show its distribution on the world map.
b) Draw a climate graph for this climate in the box provided above.
c) Complete the other boxes for your chosen climate.

Now test yourself

Use information from the climate graph to describe the distributions of temperature and precipitation for the hot semi-arid climate.

TESTED

Typical weather of UK high- and low-pressure weather systems

An Atlantic depression

Depressions (low-pressure systems) are a familiar part of the UK **weather** scene. Many develop in the western Atlantic and travel across the ocean before passing over western Europe. In the northern hemisphere, air spirals in an anti-clockwise direction towards its centre.

> **Weather** Day-to-day changes in atmospheric conditions
>
> **Warm front** A surface divide between a cold and warm air mass; temperatures rise as the front passes over
>
> **Cold front** A surface divide between a warm and cold air mass; temperatures fall as it passes over
>
> **Occluded front** A front in which the divide between warm and cold air masses is above the ground. There is little temperature change as the front passes over

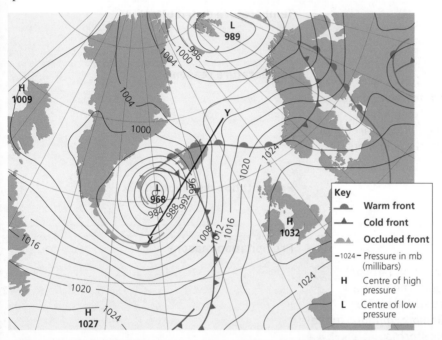

Key
- Warm front
- Cold front
- Occluded front
- −1024− Pressure in mb (millibars)
- H Centre of high pressure
- L Centre of low pressure

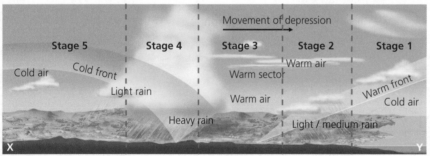

Figure 4 Weather map and section showing a deep area of low pressure in the north Atlantic Ocean on 4 September 2003.

Now test yourself

Draw a sketch cross-section through an occluded front.

Revision activity

1 Copy and complete the table below to show how the weather pattern changes from point Y to point X.

Stage on section	Temperature	Cloud cover and type	Precipitation	Wind speed and direction
1				
2				
3				
4				
5				

Anticyclones

Anticyclones (high-pressure systems) don't show the same pattern of progression as depressions. They are quite static air masses that bring their own set of weather conditions, often for long periods of time. In the northern hemisphere, air spirals outwards from the centre in a clockwise direction. The **isobars** are far apart and winds are weak.

> **Isobars** Lines connecting points of equal air pressure. The closer together they are the stronger the wind
>
> **Convectional rain** Precipitation from thunderstorms caused by air rising over a hot land surface

Now test yourself

Add the following information to Figure 4:
a) labels: depression, anticyclone, rising air, falling air
b) arrows to show the wind circulation.

Contrasting seasons

There is a huge difference in the summer and winter weather conditions brought to the UK by anticyclones.

Summer	Winter
Hot or warm days, warm nights	Cool days and cold nights
Clear skies and Sun high in the sky	Clear skies and Sun low in the sky
Chance of **convectional rain**	Frost and fog formation

It's different in the tropics

Rarely do extreme storms affect the UK. However, people living closer to the Equator than we do often suffer from life-threatening conditions brought by both high- and low-pressure systems. In recent years, many areas that experience high pressure have been doing so for longer periods of time. This results in drought. On the other hand, there seems to be an increase in the number of intense low-pressure systems that create damaging winds and heavy rainfall that causes flooding.

There is more on drought and desertification on page 111. Intense low-pressure systems (cyclones) are further explored on page 79 and their tracks are shown in Figure 1 on page 111.

Exam practice

Look at Figure 1 on page 111. Describe the distribution of areas where drought is a problem. [2] [4 lines]

Two contrasting extreme weather events

You have studied an extreme high-pressure event outside the UK and a similar low-pressure event, also outside the UK. Use these pages to organise your knowledge.

An extreme high-pressure event

We have already examined the causes of the very stable conditions brought by anticyclones. They are the same for these extreme examples. These, though, are the result of a sequence of events that result from intense heating at the Equator.

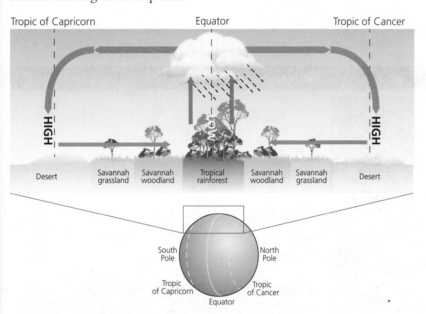

Figure 5 **How solar heating at the Equator creates the intertropical convergence zone.**

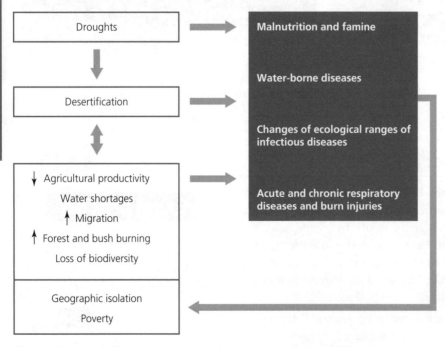

Figure 6 **Some effects of prolonged drought and desertification.**

An extreme low-pressure event

There are two main causes of low-pressure events in tropical regions.

Monsoons

A **monsoon** is the result of a reversal of high and low pressure between summer and winter. There are two main monsoons: one affecting west Africa and one affecting Australia and South-East Asia. Monsoon winds blowing on to land from sea areas can cause severe flooding.

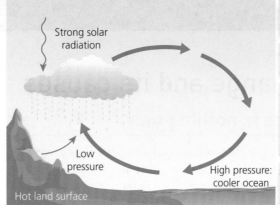

1 The ground is strongly heated by solar energy.

2 The air rises, creating a zone of low pressure.

3 Moist air from above the Indian Ocean is drawn in to the area of low pressure to fill the gap.

4 The moisture condenses, forming towering rain clouds.

Strong solar radiation

Low pressure

High pressure: cooler ocean

Hot land surface

Figure 7 Circulation of the atmosphere over South Asia during July.

Now test yourself

Figure 7 shows the processes that lead to very high rainfall in South Asia. Describe the sequence of events that will lead to very dry conditions over the same area in January.

Monsoon A wind that reverses its direction from winter to summer

Cyclones Intense low-pressure systems. Known as hurricanes and cyclones in different parts of the world

Cyclones

Cyclones have their origins in ocean areas when water temperatures reach 27 °C. This triggers a sequence of events:

The air above is heated → the moist air rapidly rises and cools → dense clouds form and heavy rain falls → the seas provide more moist air → the sequence repeats.

As with all low-pressure systems, the circulation of air is anticlockwise in the northern hemisphere and clockwise in the southern hemisphere. Isobars are extremely close together. Wind speeds are at least 75 mph and have been greater than 200 mph. Both flood and wind damage can be huge in places experiencing a cyclone.

Revision activity

Create a revision card for an extreme low-pressure event. Organise your work under the headings of 'Causes', 'Effects on different groups of people' and 'Responses'.

Responses of high-income countries (HICs) and low-income countries (LICs) to cyclones

Ability to respond	HIC	LIC
Prediction and monitoring of the track of the event, its intensity and potential to damage property. Access to satellite technology to do so rapidly	High	Low
Training of emergency services and their ability to respond speedily when a cyclone hits	High	Low
Educating people on how to respond in the event of a cyclone. Includes the ability of people to read printed advice	High	Low
The ability of individuals to prepare for an event: to protect property and to store emergency food, water and other items	High	Low
Development of infrastructure (transport and communications) that is resistant to the event	High	Low
Building dwellings that are capable of standing up to the strong winds	High	Low
The ability to use national government funds to respond immediately to provide food and shelter to affected people and, in the longer term, to repair the damage	High	Low
A reliance on aid from outside the country	Low	High

Climate change and its causes

Climate change is nothing new

During the **Quaternary** period, air temperatures have fluctuated but the main trend has been towards colder conditions. Within that time, though, there have been **glacial periods** separated by **inter-glacial periods**. A core sample of ice taken in 1950 showed the pattern indicated in Figure 1.

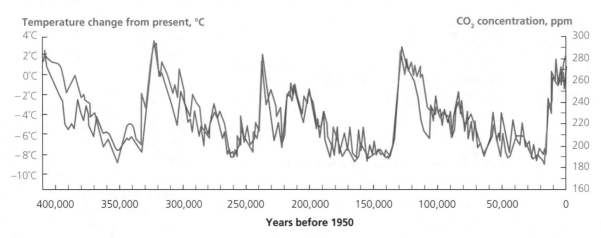

Figure 1 Temperature and CO_2 concentration (ppm) in the atmosphere over the past 420,000 years.

Now test yourself

Label the graph with a 'G' where you think there has been a glacial period and an 'I' for each inter-glacial period.

Research suggests that the temperature changes are caused by movement of the Earth. The Earth wobbles a little bit on its axis. This affects its tilt in relation to the Sun. The Earth's orbit around the Sun is also not circular but more egg-shaped. The effects on the Earth's temperature of these movements can be calculated. Some scientists predict that we should move into a new glacial period in the next 50,000 years.

Quaternary The geological period of time stretching from about 2.6 million years ago to the present day

Glacial periods Times of colder temperatures when greater areas of land are covered by ice

Inter-glacial periods Warmer times when the ice cover retreats

The greenhouse effect

Whatever the stage in the cycle described above, radiation from the Sun enters the Earth's atmosphere. The sequence that happens after that results in a trapping of heat known as the greenhouse effect.

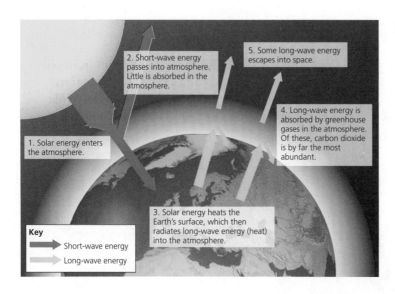

Figure 2 The greenhouse effect.

2. Short-wave energy passes into atmosphere. Little is absorbed in the atmosphere.

5. Some long-wave energy escapes into space.

1. Solar energy enters the atmosphere.

4. Long-wave energy is absorbed by greenhouse gases in the atmosphere. Of these, carbon dioxide is by far the most abundant.

3. Solar energy heats the Earth's surface, which then radiates long-wave energy (heat) into the atmosphere.

Key
→ Short-wave energy
→ Long-wave energy

Human interference

REVISED

The amount of carbon dioxide (CO_2) in the atmosphere is a key factor in how much heat is retained by the greenhouse effect. The activities of humans affect this. Many scientists believe that the enhanced global warming caused by these activities is enough to break out of the natural fluctuations seen in Figure 1.

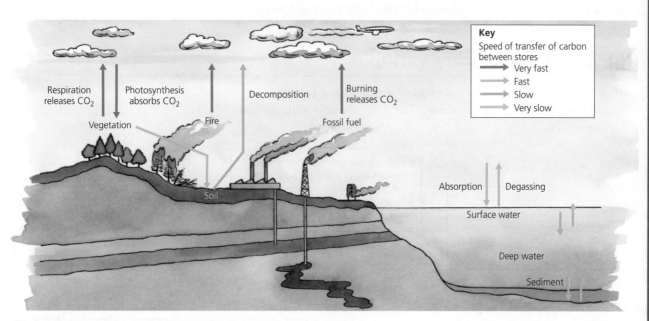

Respiration releases CO_2

Photosynthesis absorbs CO_2

Vegetation

Fire

Decomposition

Burning releases CO_2

Fossil fuel

Soil

Key
Speed of transfer of carbon between stores
→ Very fast
→ Fast
→ Slow
→ Very slow

Absorption   Degassing

Surface water

Deep water

Sediment

Figure 3 The carbon cycle, showing fast and slow transfers.

Revision activity

1 Look at Figure 1. Compare the pattern of temperature change and CO_2 concentration in the atmosphere.
2 Draw a sketch graph to show predicted changes for the next 50,000 years: a) without enhanced global warming and b) with enhanced global warming.
3 Look at Figure 3. Make a list of carbon transfers that are natural and another of transfers that are caused by people. Colour each according to their speed of transfer.

Exam practice

'Global warming is the result of human activity.' To what extent do you agree with this statement? Justify your decision.   [8] [1¼ pages]

ONLINE

The consequences of climate change

Climate change is not only the increase in global temperatures we are currently experiencing. The shifts in temperature cause the whole pattern of global air circulation to alter. This means that rainfall patterns and the intensity of weather events will sometimes be significantly different from our usual weather. The World Health Organisation (WHO) published a map to show the worldwide distribution of deaths due to climate change.

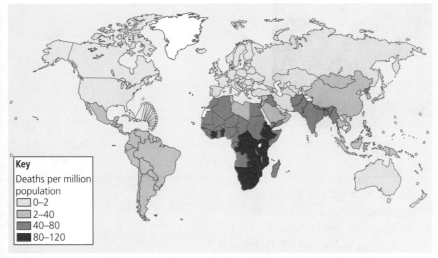

Key
Deaths per million population
☐ 0–2
▨ 2–40
▨ 40–80
■ 80–120

Revision activity

The map shows figures for 2000. Use the internet to find a more recent map. How has the situation changed?

Figure 4 Estimated deaths per million of population attributed to climate change in the year 2000. Source: WHO World Health Report 2002.

The consequences are varied

REVISED

Climate change affects both the activities of people and processes operating in the natural environment. The diagram below shows some of the main effects.

2 Habitats will alter, with some plant species dying out to be replaced by others. Broken food chains will result in some wildlife migrating or dying out

1 Changing rainfall and temperature patterns. Some farming areas will no longer be productive. In other areas farmers will need to change their crops and animals

3 Providing drinking water will become more difficult in some areas. Other areas will have too much water, causing flooding

Consequences of climate change

4 People may be forced out of the areas in which they live as conditions become more difficult. They will try to migrate to other areas

5 As weather conditions change, some places will become unpopular as holiday destinations. Others will become more popular. Employment opportunities will change

Figure 5 The consequences of climate change.

Revision activity

Produce a revision card for two of the effects shown in Figure 5. Each card should contain:

- a sketch map to show the location(s) of the place(s) affected
- specific causes
- specific and detailed effects on people and/or the environment.

The effects of climate change on coastal communities

Global warming reduces the amount of water stored as a solid on land. This loss of 'white' ice and snow reduces the amount of reflected heat. Temperatures then rise more. As liquid water heats it also expands. This increases the volume of water in our seas and oceans. So, sea levels rise. In some places, the sea-level rise will be accompanied by more severe storms than at present. This all contributes to increasing risks of coastal flooding.

Contrasting management challenges

As with low-pressure weather events (page 79), the challenges faced by a country to cope with the effects of rising sea levels and coastal flooding will differ depending on how much of the country is at risk and on the wealth of the country.

Potential challenges are shown below.

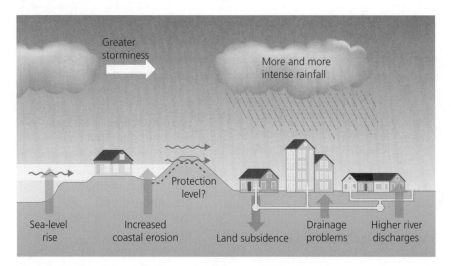

Figure 6 **Possible challenges facing some UK coastal areas.**

Revision activity

You have studied the possible impacts of rises in sea level and increased frequency of storms on coastal communities in two countries. They are at different levels of economic development. For each of these, complete a large copy of the table below.

Country	Location of community
Potential impact of climate change	
Specific information about the areas affected and the possible effects on people and the environment	
Management challenges	
Specific detail of the strategies required to counter the effects described above	
Responses to the challenges	
The reality of what management strategies the country's government is likely to be able to use	

Explain any differences in response between the two countries.

Climate change: reducing the risk

Addressing climate change

Since 1990, there have been international conferences organised by the United Nations aimed at reducing climate change. Some have made clear progress:

- 2005: Kyoto Protocol – an agreement to reduce greenhouse gas emissions negotiated in 1997.
- 2012: Doha Amendment – limit global temperature increases to below 2 °C until at least 2020.
- 2015: Paris Agreement – aim to reduce increases to 1.5 °C this century and for richer countries to provide help to poorer ones.

Maldives

The Republic of Maldives is a low-lying country of 1190 islands in the Indian Ocean. It has a population of 350,000. Its greatest income is from tourism. A rise in sea level of 0.5 m by 2100 would drown 77 per cent of the land area. A rise of 1 m would make the islands uninhabitable by 2085.

India

India, an NIC, has a growing manufacturing sector that burns fossil fuels. It also uses them to produce more electricity to improve upon the 55 per cent of its population that have access to the grid. Feeding its 1.3 billion population contributes methane (a greenhouse gas) from livestock rearing and rice growing.

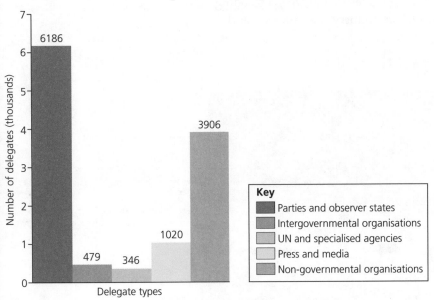

Figure 7 Types of delegates attending the United Nations Climate Change Conferences.

It is never easy reaching agreement between representatives of different countries. Some are richer than others. Some rely more on manufacturing that creates greenhouse gases than others. Some have already developed 'clean' technology. Others have not. Even when a country's government has signed an agreement, there may be a change of group governing the country. This new government may opt out of the agreement.

Exam practice

Why might countries like the Maldives and India have different attitudes to climate change? Use only evidence from the boxes above to support your answer.   [3] [5 lines]

Revision activity

Look at Figure 7. Choose four different groups of delegates who attend the conference. Produce a table to show the different views they would bring to the conference. Why would they hold each of their views? The table has been started for you below.

Delegate	Views	Reasons for views
1 Name:		

Why the UK needs to respond to climate change

REVISED

A UK government climate risk assessment group suggests a number of direct effects of climate change on the UK:

- increased flooding from heavy downpours
- extremely wet winters are five times more likely
- flood risk from rising sea levels
- a rise from 330,000 to 1.2 million houses in danger of flooding by 2080
- increased summer heatwaves put pressure on health services
- at least 25 million people with water supply deficits by the 2050s.

These negative effects are partly balanced by possibly lower deaths from winter cold, new crops being able to grow here (provided there's enough water) and wider tourism opportunities.

What can the government and individuals do?

REVISED

Apart from contributing to international conferences, the government can use both laws and advice related to:

- the electricity supply
- the energy efficiency of new and existing buildings
- fuel efficiency and emissions of vehicles.

The government can also invest money in these and other strategies like the investigation of low-carbon energy sources.

Each local council is legally obliged to publish a policy statement, a climate change adaptation action plan. Each plan shows how the council will manage the predicted effects of climate change in its area.

In this top-down model we finally reach those decisions you and I might make to help reduce climate change. These are most important to us in that they are ones over which we have the greatest control.

> **Revision activity**
>
> 1 Complete the table below. How many of these activities do you and your family follow?
> 2 Use the internet to look up your local council's climate change adaptation action plan. What are its main elements?

Individual and family strategies	Contribution to tackling climate change
Buy local produce	
Use energy-efficient electrical equipment	
Buy from renewable energy companies	
Insulate the house	
Buy durable goods	
Pack the fridge tightly	
Walk or cycle more often	
Eat less meat	
Use public transport	
Take own carrier bags when shopping	
Take holidays in the UK	
Save and use leftover food	
Boycott tropical hardwood furniture	
Buy a low-emission car	

How ecosystems function

A complex interrelationship

Each biome has its own unique character. This is mainly dependent on the climatic conditions within that biome. The plants that survive in that climate, the soils they grow in and the animals, from **decomposers** to humans, that feed within the biome are all interrelated. A biome is an extremely complex mix of **biotic** and **abiotic** elements.

The major **flows** of **nutrients** between the **stores** are shown in Figure 1. **Energy** is required by plants to convert nutrients into plant tissue. The light energy of the Sun is converted into chemical energy by plants. This then flows through the biome as animals eat plants and are, in turn, eaten themselves.

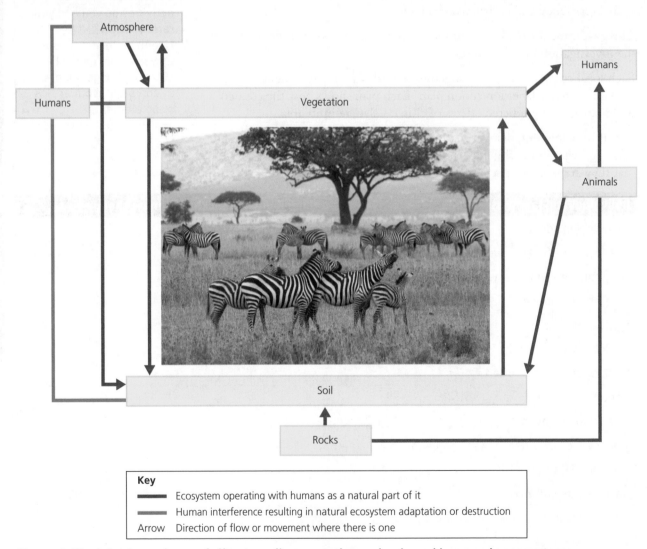

Figure 1 The interdependence of climate, soils, vegetation, animals and humans in ecosystems.

Revision activity

1 Write the number of each of the following statements at a suitable point on Figure 1.

1	The activities of people living outside the ecosystem may create conditions that lead to changes in the climate on which the ecosystem depends
2	Decaying animals and plant life will create acids that help break down the rock and dissolve minerals in it to act as nutrients
3	People may chop down naturally occurring trees and other plants for commercial reasons without replacing them
4	A source of water is necessary to dissolve nutrients for them to be taken into the plants. The speed at which soil material is weathered into a dissolvable form depends on soil temperature
5	People living as part of the ecosystem will gather fruit, nuts and other naturally occurring plant food
6	Fallen leaves and twigs will return nutrients to the soil, as will the roots and trunks of plants that have died
7	Nutrients are transferred from the soil into the plants through their roots. At least seventeen different minerals are required for plant life and almost all of these come from the soil
8	Animals return nutrients to the soil daily through natural processes. When dead, their decomposing bodies also act as sources of nutrients. Animals like worms also help mix the soil through their activities
9	The hunting of animals for food is a feature of people living as an integral part of an ecosystem
10	While not all animals eat plants in an ecosystem, those that don't *do* eat the animals that eat vegetation. Animal life in any ecosystem would not survive without the plant life
11	Carbon and oxygen are taken into plants through their leaves and are an essential part of building new plant cells through photosynthesis
12	Such activities as the extraction of minerals by open-cast methods and the drowning of land for reservoirs will result in the removal or destruction of soil
13	The rocks on which a soil rests have a large influence on an ecosystem. Broken-down rock gets mixed with the material from plant and animal life to create the soil
14	Photosynthesis creates oxygen which is returned to the atmosphere through the plant's leaves
15	Some nutrients dissolved in water are washed downwards out of the soil in a process called leaching

2 Add a label to show where 'decomposers' are active.

Decomposers Mainly bacteria and fungi in the soil that break down dead flora and fauna

Biotic Components of an ecosystem that are living or have once lived

Abiotic Non-living components of an ecosystem

Nutrients The chemicals that living members of ecosystems require for growth

Flows Movements of nutrients and energy through the ecosystem

Stores Places where nutrients and energy are at rest

Energy The component of the ecosystem that comes from the Sun

The global distribution of biomes

Ecosystems vary in size and can be found:

- on a small scale, for example a local pond
- on a global scale, for example the tropical rainforest **biome**.

A local woodland or pond has its own collection of **flora** and **fauna** that depend on each other for their existence and operate within the same annual conditions of rainfall and temperature. The character of each ecosystem is mainly a response to the area's climate. The distribution of each biome will, therefore, loosely match distinct climate zones.

> **Ecosystem** A community of flora and fauna and the environment in which they live
>
> **Biome** An ecosystem at a global scale
>
> **Flora** Plant life
>
> **Fauna** Animal life

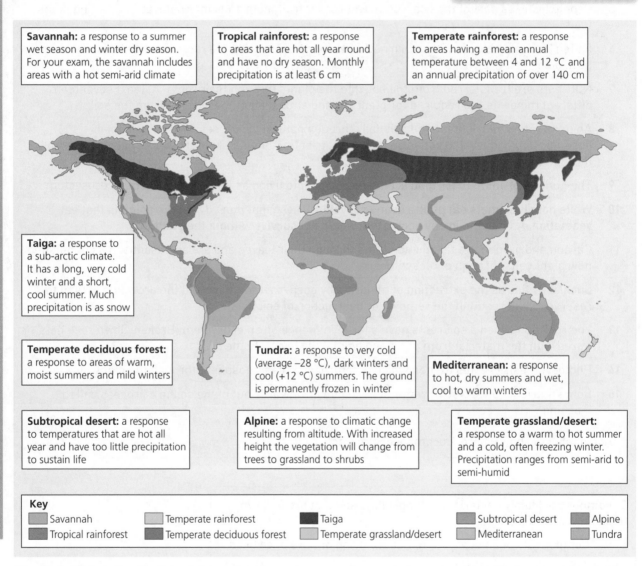

Savannah: a response to a summer wet season and winter dry season. For your exam, the savannah includes areas with a hot semi-arid climate

Tropical rainforest: a response to areas that are hot all year round and have no dry season. Monthly precipitation is at least 6 cm

Temperate rainforest: a response to areas having a mean annual temperature between 4 and 12 °C and an annual precipitation of over 140 cm

Taiga: a response to a sub-arctic climate. It has a long, very cold winter and a short, cool summer. Much precipitation is as snow

Temperate deciduous forest: a response to areas of warm, moist summers and mild winters

Tundra: a response to very cold (average –28 °C), dark winters and cool (+12 °C) summers. The ground is permanently frozen in winter

Mediterranean: a response to hot, dry summers and wet, cool to warm winters

Subtropical desert: a response to temperatures that are hot all year and have too little precipitation to sustain life

Alpine: a response to climatic change resulting from altitude. With increased height the vegetation will change from trees to grassland to shrubs

Temperate grassland/desert: a response to a warm to hot summer and a cold, often freezing winter. Precipitation ranges from semi-arid to semi-humid

Key
- Savannah
- Tropical rainforest
- Temperate rainforest
- Temperate deciduous forest
- Taiga
- Temperate grassland/desert
- Subtropical desert
- Mediterranean
- Alpine
- Tundra

Figure 2 Global distribution of the major biomes.

The hot semi-arid biome in detail

A food web

A food web is just a more detailed view of the 'vegetation → animal' link shown in Figure 1 on page 86. Everything is dependent on the **producers** that grow in response to the climate. Each upward link on the web shows the energy flow from one group of consumers to another. The direction of energy flow is shown by the position of each arrowhead. Within each food web it is possible to see individual food chains. Each of these is an upward route through the food web involving one species in each category from **primary producer** to **tertiary consumer**. In many food webs you will also find **omnivores**.

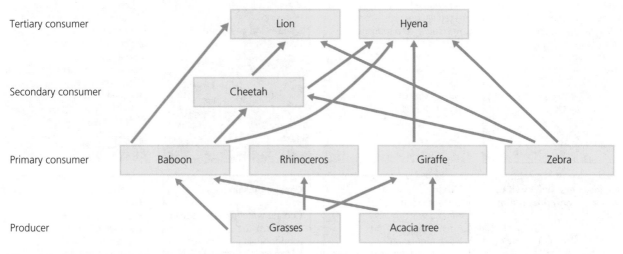

Figure 3 Food web for the semi-arid grassland ecosystem.

Exam practice

Study the food web above.
a) Complete the food web to show that: zebras eat grasses. [2]
b) Complete the following food chain:

grasses → _____ → _____ →

_____. [2]

c) Explain why the food web would alter if acacia trees were to die out. [6] [12 lines]

Producer Plant life that is the source of energy for the animals that eat it

Primary consumers Animals that eat plants; also called herbivores

Secondary and tertiary consumers Progressive stages of animals eating other animals; also called carnivores

Omnivores Animals that eat both plants and animals

Small-scale ecosystems in the UK

An ecosystems overview

Ecosystems are found in both urban and rural areas.

Almost all rural ecosystems in the UK are actually the result of some form of influence by people. Farmland is managed and even the more isolated parts of our country have a character that is modified by human activity. For example, the grassy slopes of mountain and moorland areas are the result of the regular feeding of farmed animals, usually sheep.

Mountains, moorlands and heaths
Food
Fibre
Fuel
Fresh water
Climate regulation
Flood regulation
Wildfire regulation
Water quality regulation
Erosion control
Recreation and tourism
Aesthetic values
Cultural heritage
Spiritual values
Education
Sense of place
Health benefits

Semi-natural grasslands
Food
Biofuels
Fresh water
Genetic resources
Climate regulation
Air and water quality regulation
Recreation and tourism
Aesthetic value
Cultural heritage
Spiritual values
Education
Sense of place
Health benefits

Enclosed farmland
Food
Biofuels
Fresh water
Climate regulation
Pollution control
Water quality regulation
Pollination
Disease and pest control
Recreation
Aesthetic values
Cultural heritage
Education
Sense of place

Woodlands
Timber and fuel wood
Fresh water species diversity
Climate regulation
Erosion control
Flood regulation
Disease and pest control
Air and water quality regulation
Soil quality regulation
Noise regulation
Recreation and tourism
Aesthetic values
Cultural heritage
Education
Employment
Sense of place

Freshwaters: open water, wetlands and floodplains
Food
Water
Peat (horticulture/fuel)
Navigation
Bioenergy
Health products
Climate regulation
Water regulation
Water quality regulation
Fire hazard regulation
Recreation and tourism
Aesthetic values
Cultural heritage
Spiritual values
Education
Health benefits

Urban
Genetic resources
Air and water quality regulation
Noise regulation
Local climate regulation
Flood regulation
Pollination
Recreation and tourism
Aesthetic values
Cultural heritage
Spiritual values
Education
Sense of place
Health benefits

Coastal margins
Food
Cooling water (nuclear power stations)
Land for military exercises
Pharmaceutical products
Wild species diversity
Coastal defence
Water quality regulation
Pollution control
Recreation and tourism
Aesthetic values
Cultural heritage
Spiritual values
Education
Sense of place
Health benefits

Marine
Food
Pharmaceutical products
Pollution control
Climate regulation
Recreation and tourism
Aesthetic values
Cultural heritage
Education
Sense of place

Figure 4 The eight broad UK habitats and the benefits they bring to people.

Each of the habitats shown in Figure 4 will result in slightly differing ecosystems depending on geographical position, climate and the geology on which the soils develop. They can vary in size from whole mountainsides to a small field, a stretch of cliff top or even a garden pond.

Exam practice

Study Figure 5 on page 91.
a) What percentage of the area of the photograph is green space?
   Underline the correct answer: 35%   55%   75%.   [1]
b) Name the ecosystem found at grid reference 055922.   [1] [1 line]
c) Describe three benefits ecosystems could bring to people in this housing area.   [3] [6 lines]

ONLINE

Revision activity

Use Figure 4 to identify a habitat you have studied. Group the benefits shown for your chosen habitat in Figure 4 according to whether they are social, economic or environmental.

For each of these three groups of benefits describe in detail the effects of the ecosystem you have studied.

Urban ecosystems

Are there really ecosystems in urban areas? Aren't all urban areas built up? Well, no! Quite large parts of our towns and cities support ecosystems. Estimates suggest that urban green space in England ranges from 23 per cent in Liverpool to 58 per cent in Newcastle, with London between these two at 38 per cent. Just as ecosystems in rural areas, they provide benefits to people. They also face pressures.

Sites	Overview	Future opportunities	Future pressures
Woodland, sites of special scientific interest (SSSIs), urban forestry, scrub	Total area of 11% of urban land in UK; 600 SSSIs are within or near urban areas	Planting of new woodlands in greenbelt areas	Climate change, invasive pests and infections like ash die-back disease
Street trees	66% are in gardens and grounds; 20% in public parks; 12% street trees	Liked by locals. They improve quality of life for locals and raise house values	Climate change. Also a drop in the local water table. Cost to councils of maintenance
Public parks and formal gardens	13% of public parks are in poor condition. Most of these are in deprived areas. Includes graveyards	Improved planting. Management direct by the community. Increased use due to health awareness	Lack of maintenance by local councils. Conflicts between users. Privatisation
Domestic gardens	Cover 13% of all urban land in the UK	New awareness of ecological planting and use of permeable paving	Future water shortage. Used as hard paving/ parking areas. Housing infill in large gardens
Green (wildlife) corridors	Hedgerows and trees along walking and cycling routes and canalsides. Develop habitats and allow wildlife to migrate	Many routes suitable for development as green corridors exist in most urban areas	Privatisation of access to routes. Their blocking by roads and other infrastructure crossing them

Source: substantially adapted from a table in 'The future of the urban environment and ecosystem services in the UK', Joe Ravetz, Centre for Urban Resilience and Energy, Manchester University, October 2015.

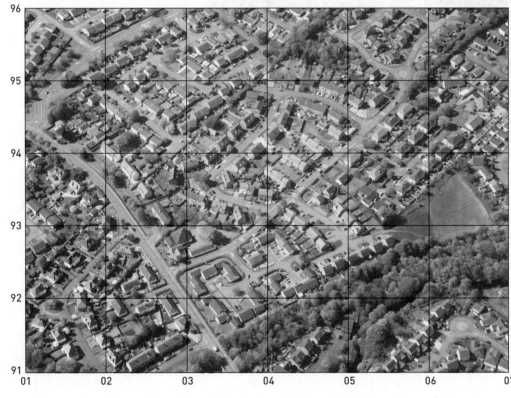

Figure 5
An aerial view of suburban housing in Glasgow.

The Attenborough Nature Reserve and its ecosystems

The Attenborough Nature Reserve is about 6 miles south-west of Nottingham. It is on the floodplain between the left bank of the River Trent and a railway line that serves the commuter village of Attenborough. It formed as a result of over 70 years of gravel extraction from the floodplain of the river. Flat-bottomed boats still pass through the reserve filled with gravel from areas that are still being worked by Cemex UK Ltd, which owns the 145-hectare site. A map of the area is on page 70.

a)

b)

c)

Figure 6 Some ecosystems within the Attenborough Nature Reserve. a) Sedges, b) a rove beetle and c) a leopard moth.

At Attenborough there are both standing and moving water ecosystems. The moving water is the River Trent and the standing water the flooded gravel workings. On the edges of these are reed beds with common reeds and sedges, the producers. These support primary consumers like the leopard moth and secondary consumers including the rove beetle.

On their landward edges, these give way to a wet woodland ecosystem of alder, birch and willow. The cranefly, soldier fly and netted carpet moth feed off these. In turn, they are eaten by frogs, toads and small birds. These provide food for ducks, geese and foxes which are themselves consumed by larger birds like hawks and owls and mammals like foxes and weasels.

Now test yourself

1 Draw a reed bed food chain involving the three species shown in Figure 6.
2 List the species found in the wet woodland ecosystem under the following headings:
  – producer
  – primary consumer
  – secondary consumer
  – tertiary consumer
  – quaternary consumer.

TESTED

The Attenborough Nature Reserve was opened in 1965 by Sir David Attenborough. He returned in 2005 to open the nature centre, which later won a gold ecotourism award. Management of the site has to balance the needs and demands of a variety of people. Because the Attenborough Nature Reserve is close to an urban area, it attracts quite large numbers of visitors. They come for different leisure and educational purposes. They place different demands on the ecosystems and are sometimes in conflict with each other. These **stakeholders** will often have different viewpoints about the ways in which the ecosystem is being managed; agreeing with some strategies while disagreeing with others.

Stakeholders A group of people who, in this case, are affected by an ecosystem and its management

- Artificial nesting banks

- Café, shop and study rooms

- Educational programmes

- Paid car park provision

- Shrub management

- Birdwatching hides

Figure 7 Twelve features in words and pictures of the Attenborough Nature Reserve.

Possible influences on stakeholder views

Stakeholder	Positive influences	Negative influences
Wildlife Trusts member	Attracts a greater variety of birds Protected breeding for rare species	Very popular with non-birdwatchers Nature Centre ruins natural 'feel'
Teacher	Provides activities for students Has indoor study facilities	Visitor centre and paths can be busy
Attenborough resident	Recreational facilities next to home Raises house prices Stores floodwaters	Visitors parking on narrow village roads Visitors intrude on peace of village
Jogger	Pleasant surroundings Paths well maintained Choice of signposted routes	Share paths with other users Car park charges

Revision activity

Either for this ecosystem or for a different small-scale ecosystem that you have studied:
a) suggest reasons for the different viewpoints of stakeholders
b) how do different views result in conflict between stakeholders and increase the challenges of managing the ecosystem?

Exam practice

'It is not possible to successfully manage an ecosystem for a sustainable future.' To what extent do you agree with this statement? Use evidence from an ecosystem you have studied. [8] [32 lines]

ONLINE

2 Ecosystems under threat

Using ecosystems for human profit (or loss)

You already know that the health of ecosystems can often be indicated by changes in their **biodiversity**. The map below shows the percentage of **indigenous species** that are now found in different parts of the world. Usually, the loss of indigenous species is the result of human activities that use and change ecosystems and environments in order to obtain food, water and energy resources.

> **Biodiversity** The number of different species found in an ecosystem
>
> **Indigenous species** Those that are the natural response to conditions of climate and rock type in an area

Threats to biodiversity within ecosystems

REVISED

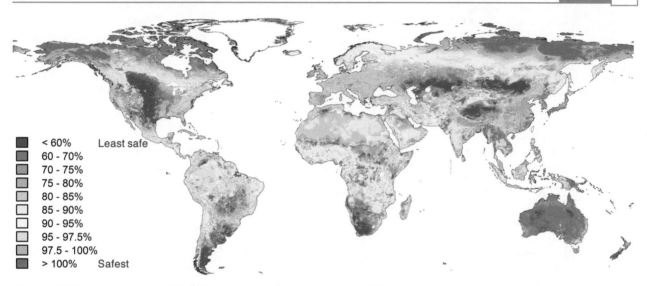

■	< 60% Least safe
■	60 - 70%
■	70 - 75%
■	75 - 80%
■	80 - 85%
□	85 - 90%
□	90 - 95%
■	95 - 97.5%
■	97.5 - 100%
■	> 100% Safest

Figure 1 The percentage of indigenous species remaining in the ecosystem.

According to the Natural History Museum, loss of biodiversity has reached unsafe levels across 58 per cent of the world's land area. There are 34 biodiversity 'hotspots' around the world. These are areas that are the only homes for large numbers of species. In 22 of these, there are unsafe levels of biodiversity loss. The museum suggests that we are reaching a situation in which people will need to step in and try to manage these ecosystems.

The main causes of loss of biodiversity

Most loss of species is the result of human activities. These include:
- destroying ecosystems to use the land for agriculture and buildings
- the introduction or invasion of plants and animals that compete with the original species
- outdoor recreation and tourism often destroying habitats and disturbing breeding patterns
- over-hunting
- atmospheric pollution, including the effects of acid rain.

> **Now test yourself**
>
> 1 Describe the distribution of areas at greatest risk of biodiversity loss.
> 2 Suggest reasons for the distribution you have described.
> 3 Why is it difficult to give the reasons for the loss of biodiversity?
>
> TESTED

Effects on humans: the gains

The short-term gains of exploiting our ecosystems can be great. Look at the photographs below. They show three common reasons for changes to ecosystems by humans: energy generation, logging and farming. Large expanses are also flooded in order to store the ever-increasing amounts of water needed by people.

Figure 2 **Three different ecosystems: a) Mojave Desert, USA; b) Borneo; c) Thailand.**

Photo	Ecosystem	Use	Human gain	Local ecosystem loss
a	Hot desert	Solar energy generation	Energy for 140,000 homes Uses less water than conventional generation methods Greatly reduces overall greenhouse gas emissions	Plants removed during construction Groundwater loss for washing mirrors Golden eagle/bighorn sheep habitats destroyed Birds killed by concentrated Sun's rays
b	Tropical rainforest	Logging	Formal employment Increased income for country Raw materials for manufacturing industry	Fragmentation of forest Damage to unlogged trees Increased soil erosion Increased sediment to river; flooding
c	Mangrove forest	Shrimp and prawn farming	Farmers offered subsidies to expand Main source of shrimps to USA, Japan and European Union Most earnings go to people from outside the local area	Destabilises coast – more flooding Loss of wild fish habitats Antibiotics and pesticides leak into ecosystem from farms

Effects on humans: the losses

Many scientists are now convinced that biodiversity loss is as great a concern for life on Earth as climate change. Within some ecosystems, people who rely on hunted animals and collected plant material will lose these supplies. Its effects will, though, also be much broader:

- Food supplies: both farm crops and animals become more vulnerable to pests and disease due to invasion of farmland by indigenous species whose habitats are under pressure.
- The insects that pollenate farm crops die out over large areas.
- The large areas of loss affect atmospheric processes as transpiration is reduced from large areas of land. Water supply to some areas becomes irregular or inadequate due to changed rainfall patterns.

Revision activity

1 Describe each of the activities shown in Figure 2.
2 Do you have more examples of how loss of biodiversity and other changes within ecosystems have negative effects on people in other parts of the world? Make a list.

Damage to an ecosystem

Ecosystem loss in the hot semi-arid grasslands

REVISED ☐

Hot semi-arid ecosystems are found across the world. The reasons that hot semi-arid ecosystems are under threat vary from region to region. Usually, the threat is the result of changes in the ways in which people use the ecosystem. Natural threats may change with time but they have mainly been in existence for many years and the ecosystem has survived them. For example, fires caused by lightning storms and by local traditional farmers encouraging grass growth that can damage young trees.

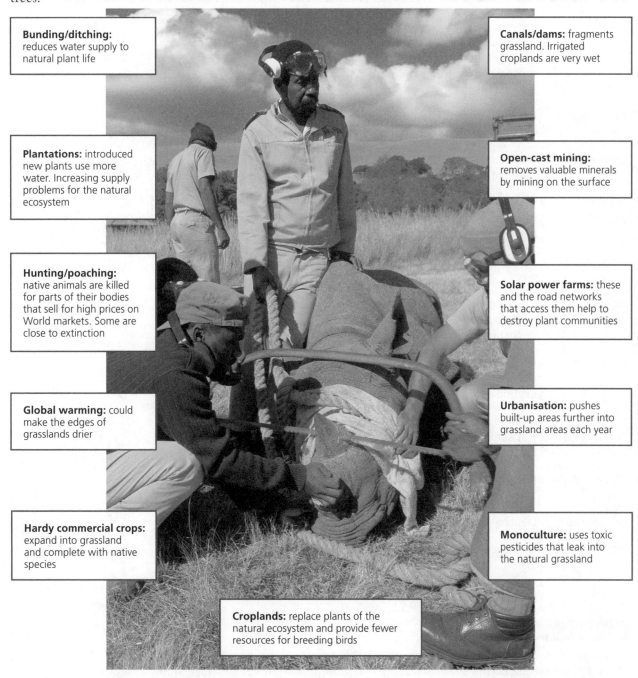

Bunding/ditching: reduces water supply to natural plant life

Plantations: introduced new plants use more water. Increasing supply problems for the natural ecosystem

Hunting/poaching: native animals are killed for parts of their bodies that sell for high prices on World markets. Some are close to extinction

Global warming: could make the edges of grasslands drier

Hardy commercial crops: expand into grassland and complete with native species

Canals/dams: fragments grassland. Irrigated croplands are very wet

Open-cast mining: removes valuable minerals by mining on the surface

Solar power farms: these and the road networks that access them help to destroy plant communities

Urbanisation: pushes built-up areas further into grassland areas each year

Monoculture: uses toxic pesticides that leak into the natural grassland

Croplands: replace plants of the natural ecosystem and provide fewer resources for breeding birds

Figure 3 How hot semi-arid grasslands are affected by human activity. For example, the photo shows conservationists removing the horn from a black rhino to discourage poachers from killing the rhino for its horn.

Two ecosystems compared

Each human activity that causes damage to an ecosystem has an effect on the processes operating in that ecosystem, including the effect of loss of biodiversity. Such interference knocks ecosystems out of balance, depletes resources of use to humans and contributes to climate change. The need for conservation is great.

The table below shows how human activity affects processes in the hot semi-arid grassland ecosystem.

Ecosystem			
Hot semi-arid:		**Other ecosystem:**	
Location:		**Location:**	
Cause	**Effect on processes**	**Cause**	**Effect on processes**
Bunding/ ditching	Topsoil dries out; there is soil erosion and increased surface runoff		
Canals/dams	Fragmentation reduces species movement. Wetter croplands disrupt nesting birds		
Plantations	Result in death of some producers with knock-on effect up the food web		
Open-cast mining	Removes soil completely, destroying the food web		
Hunting/ poaching	Removes one consumer from the food web. Less food at the next highest level		
Solar power farms	Remove producers from the food web		
Global warming	Places greater water pressure on grasslands, leading to desertification		
Urbanisation	Removes areas of ecosystem. Greater surface runoff and reduced groundwater		
Hardy commercial crops	Change producers in the food web. Put some consumers at risk		
Monoculture	Kills some producers. Transfers some toxins into primary consumers		
Croplands	Destroy habitats of some consumers		
Burning	Improves growth rate of some producers, increasing food for consumers. Kills other producers. Change of balance in food web		

Revision activity

1 Complete the table above with information about your second ecosystem.
2 Add detail on a separate sheet about the hot semi-arid area that you have studied.
3 To what extent are the effects demonstrated at the local and the global scale? Colour code each activity to help you answer this question.

Now test yourself and exam practice answers at **www.hoddereducation.co.uk/myrevisionnotes**

Conservation and ecosystem management at Ngorongoro

REVISED

The Ngorongoro Conservation Area

This area is an attempt to manage an area of semi-arid grassland in such a way as to balance the needs of the natural environment and those of people who rely on it for their ways of life. The Maasai people are traditionally semi-nomadic herders. They move seasonally around the area with their cattle, getting the best out of the grassland. This way of life is threatened by the demands of tourism. The area is managed by the Ngorongoro Conservation Area Authority (NCAA).

Bunding Creating low earth banks to help retain water on farmers' fields

Monoculture Growing of large areas of a commercial crop

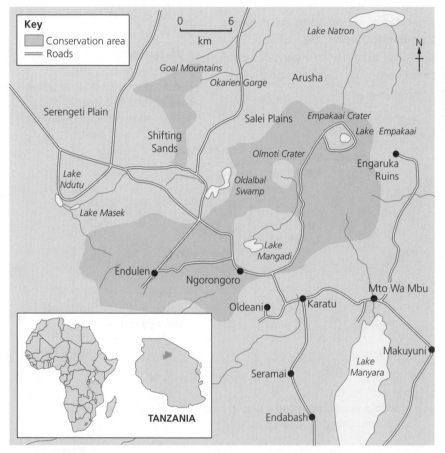

Figure 4 **The location of the Ngorongoro Conservation Area.**

Figure 5 **Maasai pastoralists.**

Exam practice

Study Figure 4.
a) Describe the location in Africa of the Ngorongoro Conservation Area. [2] [3 lines]
b) Why might tourists be attracted to this conservation area? Use map evidence only. [3] [5 lines]
c) Suggest two ways in which Figure 4 could be improved to help you answer questions a) and b). [4] [6 lines]

ONLINE

General management plan

There are three main management objectives of the NCAA:
● to conserve the natural environment
● to protect the interests of the Maasai pastoralists
● to promote tourism.

To do this, the NCAA needs to understand the balance needed to satisfy all the demands on the area. The archaeological sites need to be protected. Tourism must be concentrated on the quality of tourism experience and not just numbers of tourist facilities. Vehicle access to popular sites needs to be restricted and measures taken to ensure that visitors do not damage the environment or disturb traditional activities.

It is important that there is a need for a joint strategy developed between all stakeholders, including the NCAA and the Maasai population.

Revision activity

For the contrasting ecosystem you have studied, make notes under the following headings:
● location of the ecosystem
● main pressures on the ecosystem
● the plan to manage the ecosystem.

3 Water resources and management

Global water supply and demand

Our water footprints

Water is essential to life: plant, animal and human. Ecosystems are, in part, responses to the amount of water that is available in a particular place. People use water directly in their day-to-day lives to drink, wash and cook, for example. We also use **embedded water**, for example, the water it takes to grow cotton and in the manufacturing processes that then turn the cotton into the clothes we wear. Each item we use contributes to our **water footprint**.

> **Embedded water** Water used by other people to make grow or articles we use
>
> **Water footprint** The amount of water used by an individual either directly or as embedded water

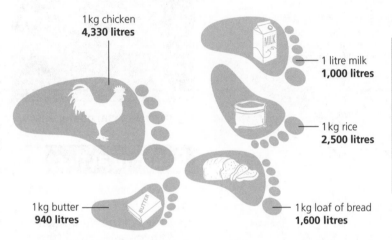

1kg chicken
4,330 litres

1 litre milk
1,000 litres

1kg rice
2,500 litres

1kg butter
940 litres

1kg loaf of bread
1,600 litres

Other footprints	Litres
T-shirt and pair of jeans	10,000
Pizza	1,260
1kg chocolate	17,000
A dozen bananas	1,920
1kg beef	13,500

Figure 1 Water footprints for selected items.

Now test yourself

1 Rank the water footprints from highest to lowest.
2 Why is this rank order of little use in working out a person's weekly embedded water usage?

Different places, different opportunities

Not all places have the same levels of access to a water supply. If there is not enough water available naturally, the cost of providing it can be expensive. Some countries have little spare water beyond that for essential uses, while others have plenty to spare for such uses as recreation in, for example, keeping sports pitches and ornamental gardens watered.

Revision activity

1 Locate and name each of the countries in Figure 2 on the world map opposite (Figure 3). Use an atlas to help you.
2 Label each country as either HIC, LIC or NIC.

Figure 2 Water use in selected countries.

Country	Annual water use by agriculture (km³/year)	Annual water use per capita (m³/person/year)	GNI per capita, PPP (current international $)*
Cambodia	2.05	158.9	2,260
Egypt	59	1,000	6,160
Ghana	0.65	49.63	1,820
India	688	615.4	3,620
Malawi	1.17	98.95	870
Niger	0.66	69.28	720
Nigeria	7.05	89.07	2,300
Pakistan	172.4	1,024	2,880
UK	0.99	171.8	35,940
USA	192.4	1,575	48,890

Water usage: the world picture

One of the most important uses of water is in providing enough food for the population. The figures don't necessarily tell the full story. For example, some countries, like the UK, import more than half of the food they consume and use a small percentage of water for agriculture. However, others, like Cambodia, import little food and use a large percentage of water for agriculture. Fourteen per cent of Cambodia's population is undernourished while the figure for the UK is about five per cent. Unlike Cambodia, the reason for undernourishment in the UK is personal diet decisions and not lack of food availability.

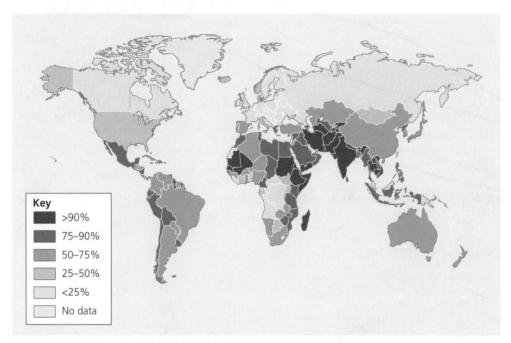

Key

■	>90%
■	75–90%
■	50–75%
■	25–50%
■	<25%
■	No data

Figure 3 The percentage of water withdrawn for use by agriculture.

Contrasts in water use

Figure 4 Lesotho: collecting water.

Figure 5 USA: irrigation.

Exam practice

Use information from the map and photos to describe the global inequality of access to water. [4] [6 lines]

Changes with time

The demand for water is dependent on two main factors:
- the numbers of people who require access to the water
- the amount of water required by each individual, either by direct access or as embedded water.

> **Water security** The extent to which a country or area is able to ensure future supplies of water to meet demand

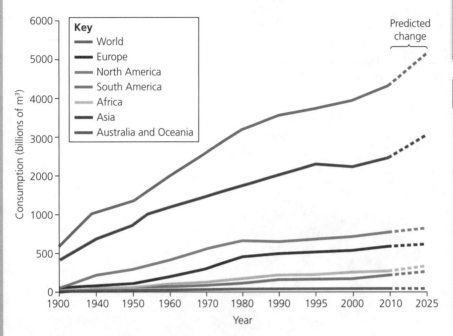

Figure 6 Global water consumption, 1900–2025.

Study Figure 7.
a) Describe the trend in population change between 1750 and 2015. Use figures in your answer. [3] [5 lines]
b) Explain why the population change is expected to level out between 2015 and 2100. [3] [5 lines]

ONLINE

Effects of consumerism

Consumerism encourages people to buy goods in ever-increasing amounts. As the global population becomes wealthier, there is a greater demand to buy and process more resources, so people's use of embedded water goes up. For example, as car ownership increases worldwide, so does water consumption. It takes almost 148,000 litres to make a car. On a smaller, but more frequently bought, scale it takes 1.4 litres of water to produce a 1-litre bottle of water. So, a combination of projected population changes and a rise in expectations are likely to have a great effect on future water demands.

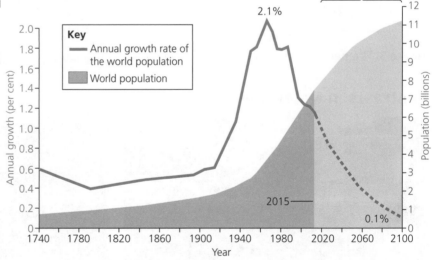

Figure 7 World population growth, 1750–2100.

Water security

Provision of water is never guaranteed. Here are some factors that may have a negative effect on future **water security**:
- climate change resulting in drought
- climate change resulting in flooding
- political change threatening supplies that cross national boundaries
- economic change threatening maintenance of expensive supplies.

Now test yourself

Suggest reasons for the differences in predicted changes in water consumption shown in Figure 6.

TESTED

Over-abstraction

Over-abstraction affects two sources of water supply:

1 Groundwater, where water is stored in porous rocks underground

> **Over-abstraction** Taking more water from a source than is capable of being replenished

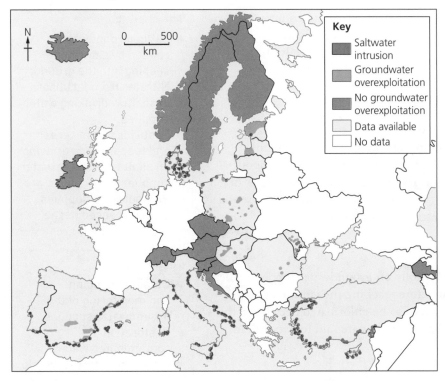

Figure 8 Groundwater over-abstraction.

There are two serious effects of over-extraction of groundwater:

● A severe drop in the water table so that water cannot be raised in sufficient quantities for an area's needs.

● In coastal areas, a lowering of the water table so that salt seawater seeps into the **aquifer** to make the stored water unsuitable for use.

> **Aquifer** An underground store of freshwater that is used by people.

2 Surface supplies such as a naturally occurring lake

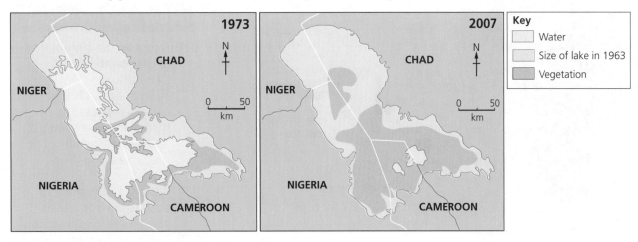

Figure 9 The shrinking of Lake Chad.

Effects of the reduction in size of Lake Chad include:

● the near death of a former fishing economy

● migration of herders creating competition for scarce resources elsewhere

● competition between communities for control of the remaining water.

The impact of over-abstraction of water

Denmark's groundwater over-extraction

Denmark is a small country in north-western Europe. It has a total area of 43,000 km² with 5.5 million inhabitants. Much of the country is underlain by chalk, a porous rock ideal for the storage of water. It is a low-lying country which makes it ideal for accessing groundwater.

Economy

Danish farmers practise intensive farming using large amounts of fertilisers, manure and pesticides. Almost 80 per cent of Denmark is farmland. In addition to cereal crops, Denmark produces roughly 25 million pigs per year as well as half a million cows that yield about 5 million m³ of milk a year. Agricultural produce is about a quarter of Denmark's exports. Denmark was the 32nd richest country in the world in 2014 according to its gross domestic product (GDP) per person.

The water supply

Water supply in Denmark is based wholly on groundwater supplies. The deep supplies need little treatment before reaching the consumer. Only those for the capital, Copenhagen, need to be chlorinated because of the distance travelled through pipes. These supplies are said to better than bottled water.

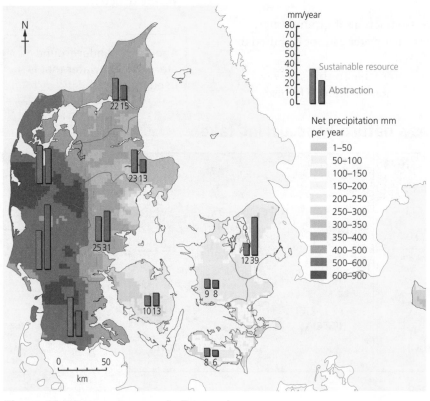

Figure 10 Water resources in Denmark.

The challenges

- Saline groundwater caused by seawater seeping into the ground.
- Nitrates from fertilisers in shallow drinking water supplies.
- Pesticides from cereal and fodder crops growing on shallow drinking water supplies.
- Pressures on supplies from increased usage.

Responses

- Detailed mapping and monitoring of the groundwater supplies.
- Restoration of contaminated waters.
- Increased afforestation to protect catchment areas.
- Regulation of the use of pesticides and fertilisers.
- Reduction of leaks in water supply pipes.
- Increased taxes on water usage.

Now test yourself

1 Describe the distribution of precipitation in Denmark.
2 To what extent is this pattern reflected in the balance between 'sustainable water resources' and their extraction?
3 Add a 'so' statement to each of the 'responses' above to help explain how Denmark is attempting to solve its water supply problems.

Lake Chad's surface water extraction

REVISED

The Lake Chad drainage basin is in west central Africa. It has an area of 2,381,635 km² and a population of over 40 million. All of these people rely on water that formerly would have flowed into the lake.

Economy

The four countries that comprise almost all of the Chad basin were among the 40 poorest countries in the world in 2014 (out of 228). The Chad economy is based mainly on **subsistence agriculture** and nomadic pastoral farming.

The water supply

Lake Chad occupies a large shallow basin towards which rivers flow from both north and south. Its drainage basin is surrounded by mountains. Hardly any water now flows into Lake Chad from most of its sources. Flow from the Chari-Logone network accounts for about 95 per cent of the water it receives, although this is less than ten per cent of its previous flow rate.

The challenges

- Reduced flow from both the Komodougou-Yobe and Chari-Logone river basins because of dam construction since 1970.
- Over-grazing and deforestation over large areas of the catchment area resulting in a drier climate.
- Use of water for both irrigation and by expanding urban areas.
- **Salinisation** of soils in the basin from too much irrigation.
- Increased migration as a result of poverty putting pressure on major urban areas.
- Political unrest and increased support for extremist groups.

Responses

Over 30 years ago, the Schiller Institute promoted the Transaqua Project to take water from the Congo River to Lake Chad. In December 2016, Powerchina, the Lake Chad Basin Commission and Nigerian authorities signed a memorandum of understanding for a study to decide whether a similar project might go ahead.

The project could:
- transfer 50 billion m³ of water a year to Lake Chad through a series of dams
- generate up to 15–25 billion kWh of hydroelectricity
- develop a series of irrigated areas for crops or livestock over an area of 50,000–70,000 km² in the Sahel
- create a huge economic zone by supporting agriculture, industries, transportation and electrical production, affecting up to twelve African countries.

> **Subsistence agriculture**
> Growing enough food for the needs of the family, with none or little left for sale
>
> **Salinisation** An increase of salt either caused by seepage into water supplies or by over-irrigation and rapid evaporation of soil water to leave salts in the soil

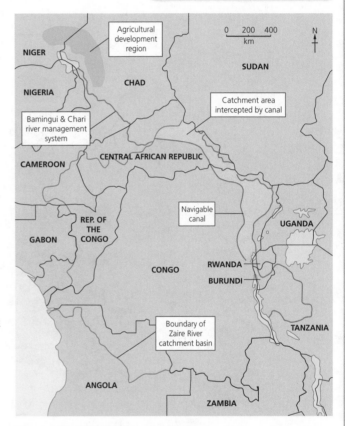

Figure 11 The Transaqua Project.

> **Revision activity**
>
> 1 Use an atlas to locate the Congo River and Lake Chad. What difficulties might be faced in transferring the water.
> 2 To what extent might the features of the Transaqua Project provide solutions to the challenges described above?

The imbalance of water supply and demand in the UK

Water does not always occur naturally where we need it most. There is a need to transfer it from places of plenty to those where the demand for water is greater than the water received. A drought is declared in an area when the supply of water is so low that normal consumption will create a risk of supplies running out.

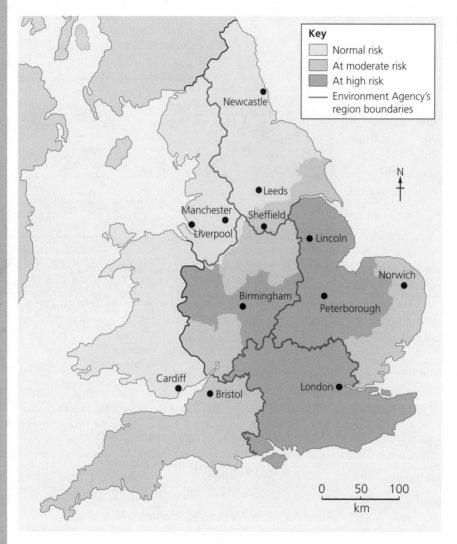

Figure 12 Drought risk in 2012 across England and Wales.

Potential pressures on drought-prone areas

REVISED

Predicted change		Effect on water balance	
A	A growing urban population	1	Increased demand for crop irrigation
B	Smaller household sizes	2	Increased demand even if individual consumption is the same
C	More intense storms	3	Increased runoff and less infiltration reduce aquifer storage
D	Drier summers	4	More water use because of smaller washing machine and dishwasher loads

Now test yourself

Draw lines in the table to connect each predicted change with its most likely effect on water balance.

TESTED

Birmingham's response

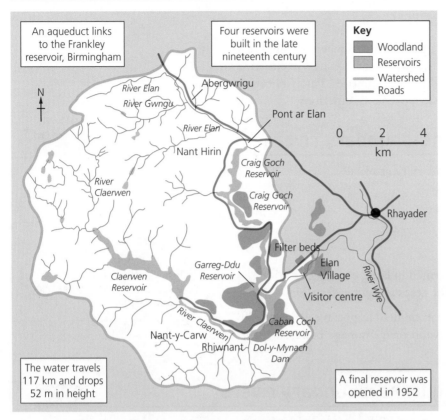

An aqueduct links to the Frankley reservoir, Birmingham

Four reservoirs were built in the late nineteenth century

Key
- Woodland
- Reservoirs
- Watershed
- Roads

River Elan
River Gwngu
Abergwrigu
Pont ar Elan
River Elan
Nant Hirin
River Claerwen
Craig Goch Reservoir
Craig Goch Reservoir
Rhayader
Filter beds
Garreg-Ddu Reservoir
Elan Village
River Wye
Visitor centre
Claerwen Reservoir
River Claerwen
Nant-y-Carw
Rhiwnant
Caban Coch Reservoir
Dol-y-Mynach Dam

0 2 4 km

The water travels 117 km and drops 52 m in height

A final reservoir was opened in 1952

Figure 13 The Elan Valley scheme provides Birmingham's water supply.

- Wooded slopes
- Steep valley sides
- Impermeable rock
- Sparse population
- High annual precipitation

Figure 14 Graig Goch reservoir and masonry dam in the Elan Valley, Powys, Wales.

From the *Birmingham Post*, 4 November 2014

Birmingham is to get a fresh source of water for the first time in over 100 years – more than a century after the Elan Valley reservoir was built in the Welsh hills to supply the city.

Severn Trent Water has announced a twenty-first-century multi-million pound scheme to construct a water pipeline linking an abstraction site on the River Severn near Stourport to Frankley Treatment Works.

The new supply project, to be completed by March 2020, has been approved to enable investment and improvement work to be carried out at the Elan Valley Aqueduct, which is showing signs of deterioration.

For over a century, Birmingham's water has been provided by the Mid-Wales aqueduct. But it is now well over 100 years old and needs increasing maintenance work.

Exam practice

In 2011, the then Mayor of London, Boris Johnson, wrote in the *Daily Telegraph*: 'Since Scotland and Wales are on the whole higher up than England, it is surely time to do the obvious: use the principle of gravity to bring surplus rain from the mountains to irrigate and refresh the breadbasket of the country in the south and east.' To what extent do you agree with him? Explain your reasons. [8] [30 lines]

ONLINE

Revision activity

Create a table to show both positive and negative effects of the Elan Valley scheme on people and the environment in the area. Use information on both pages to help you.

Revision activity

Create a revision card with the title: 'A local scale example of water transfer'.
Use the following sub-headings to organise your notes:
- supply of water
- demand for water
- meeting future demands.

Water management across national boundaries

There are differences of opinion within a country when it comes to managing water. However, these are few when compared to situations where rivers flow through several different countries. Partly because of their size, such trans-boundary rivers have many functions.

Features of large trans-boundary rivers

For several countries, a trans-boundary river usually:
● provides water for drinking
● supplies water for industry
● acts as a transport route
● produces **hydropower**
● is a source of irrigation water
● contains fish, often caught or farmed in large quantities
● provides seasonal floodwaters for crops and silt as fertiliser.

In short, trans-boundary rivers are extremely important economically and socially to most people who live in their drainage basins, whichever country they happen to live in.

> **Hydropower** Electricity generated by river water passing through turbines. Also called hydro-electric power (HEP)

Potential issues along trans-boundary rivers

Not all of the issues result from direct use of the river by the countries in its catchment. Some may result from global climate change. Those may be even more difficult to manage without a great deal of international cooperation beyond the catchment area of the river basin.

Issue	Cause	L/G
A reduction in the amount of floodwater received	Evening out of flow by damming upriver	
Disruption of river trade and travel	Obstructing river by dam construction	
Reduction in fish catches along river banks	Pollution of river waters from industrial sites	
Increased flooding of lower course of river	Sea-level rises	
Reductions in supplies of river drinking water	Increased water loss due to increased irrigation	
Reduced quality of drinking water supplies	Pollution from industry and fertilisers/pesticides	
Flooding caused by overfull reservoirs	Increased seasonal rainfall in catchment area	

Now test yourself

1 Place an L or G in the right-hand column of the table above to show whether each might be a locally or an internationally caused issue.
2 Why might it be difficult to get international agreement about how a river is used?

Managing the Mekong River

The Mekong is the twelfth longest river in the world. Its drainage basin is shared by six countries. Since 1996, these countries have worked together through the Mekong River Commission (MRC) to manage use of the river. Its 2016–20 development priorities recognise that the Lower Mekong Basin 'is home for 65 million people, 80 per cent of whom live in rural areas dependent on agricultural livelihoods. Many are poor. But all countries are expected to have reached middle-income status by 2030.'

Revision activity

1 Read the information about 'Development potential' and 'Development principles'. Place the principles in rank order according to what you think of their importance. Explain your chosen order.

2 Look back at page 67. In what ways might the creation of dams along the Mekong River affect people living in its delta area? Organise them according to whether they are positive or negative effects.

1 Current situation

- not enough water is stored to distribute it between wet and dry seasons
- little groundwater is used in the river basin
- dry-season farming is limited by low discharge rates
- only ten per cent of HEP potential is being used
- response to major floods is mainly by soft engineering methods
- water quality is good except in the heavily populated delta
- forests are being destroyed by logging and other land-use demands.

2 Development potential

- further HEP production on tributaries
- expansion and intensification of irrigated food production
- some further HEP production on the Mekong itself
- development of fisheries, navigation, drought and flood management, tourism.

3 Development principles

- environment/ecosystem protection
- equality between all countries involved
- maintain water flows on the main river
- remove any harmful effects
- maintain freedom of navigating the river
- respond to emergencies.

Agreed development priorities 2016–20

The agreed development strategies under this **five-year plan** are an extension of those for the previous one from 2011 to 2015.

- To address opportunities and threats of current developments, including closer cooperation with China about its HEP dams. This should improve dry-season flow, address sediment rate change issues and provide early flood warning.
- To expand and intensify irrigated farming, improve food security and increase employment. Introduce improved seed varieties and farm practices. Also put in place drought strategies to ensure greater reliability of water.
- To improve the sustainability of HEP development by identifying ecosystems needing protection and minimising the negative social effects of damming rivers.
- To acquire knowledge about current uncertainties and to identify development opportunities. Conduct research into the impacts of changed sediment flows on erosion and deposition, delta formation, habitats and fisheries.

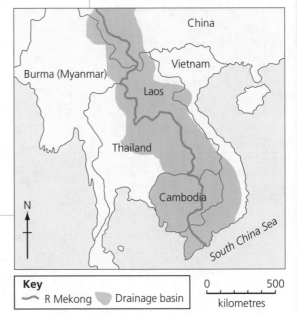

Key
~ R Mekong   Drainage basin

0   500
kilometres

Figure 15 **The location of the Mekong River and its river basin.**

Five-year plan A means of implementing development aims that have a short timeline. This enables people to measure their progress and be held accountable for their success or failure.

4 Desertification

The extent of desertification

Areas at risk of desertification

The world's desert areas are those that typically have less than 250 mm of precipitation a year. They are areas where evaporation and transpiration (evapotranspiration) greatly reduce the small amount of total precipitation to amounts that can sustain few life forms.

Many of the margins of desert areas are at risk of **desertification**. Over 1 billion people currently depend on these areas, managing to survive usually by subsistence farming. There is a contrast, though, between the effects of desertification in high-income countries (HICs) and low-income countries (LICs).

> **Desertification** The process by which fertile land becomes desert

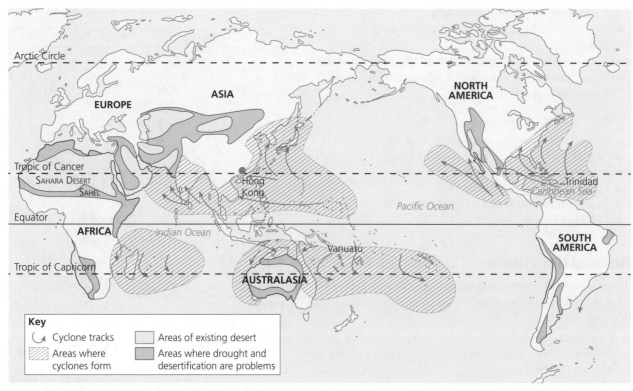

Figure 1 Regions at risk of desertification.

Revision activity

1 Label Figure 1 with the name of each desert area. Use an atlas or the internet to help.
2 a) Identify one area on the map that you have studied where desertification could cause major problems for people and another where it may cause few problems.
  b) Why did you choose each of these areas?

Now test yourself

Most of the world's hot deserts develop in areas that are dominated by high-pressure systems. Look back at pages 77 and 78.
1 Why are many of the world's hot desert areas located either side of the two tropics?
2 Why does the Sahel have wet and dry seasons?

Unreliable rainfall

One of the reasons for desertification is unreliable rainfall. Precipitation rates will differ from year to year across much of the world. It is when less than 'normal' rain falls for a large number of successive years that major problems result:

● Less water is available year on year for farming so more is taken from wells. This reduces the amount stored in aquifers.

● Less water is available to infiltrate the ground so the water taken from aquifers is not replaced.

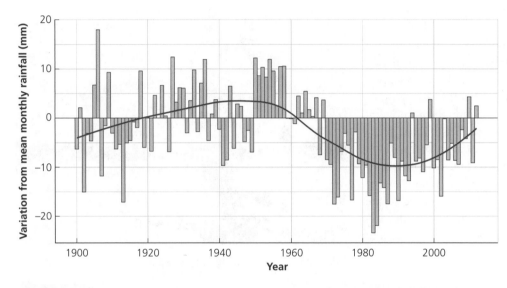

Figure 2 Variation from the mean monthly rainfall in Sahel countries 1900–2013. Each bar represents whether the average rainfall in each month of the rainy season was above or below average (when compared to the average rainfall during the period 1950–79). The line shows the trend.

Figure 3 Ploughing of traditional grasslands for crop growing.

Exam practice

a) Describe the location of the Sahel. [2] [3 lines]

b) Describe the trend in mean monthly rainfall for the rainy season between 1900 and 2013. [3]

c) Explain two effects of this difference in mean annual rainfall on people living in the Sahel. Use evidence from the photograph. [4] [6 lines]

ONLINE

Causes of desertification

Causes of desertification are quite complex. Some might be considered 'physical', resulting from changing climate patterns at a global scale. Others may be the result of the activities of people at the local and regional scales. As we know already, the global changes may also be the result of people's activities.

> **Sheet erosion** The removal by rainfall of thin layers of soil

Small-scale changes

The removal of trees and overgrazing and ploughing of land cause small-scale changes in vegetation cover. Such poor land management can have a number of effects. Tree removal will cause small areas of land to lose their shade. Soil temperature will rise and the soil will dry out further. There will also be a reduction in overall evapotranspiration resulting in less chance of convectional rain. The unprotected soil will be open to greater **sheet erosion** when there are rainstorms. Such microclimate changes can speed up desertification. Taking too much water from wells is unsustainable and, as wells dry up, people are forced to move, leaving more land to become desert.

Figure 4 **A desertified landscape.**

Changing climate patterns

Climate cause	Effect	Specific place detail
Many areas experiencing desertification are in the intertropical convergence zone (ITCZ). The position of the ITCZ migrates according to where the Sun is directly overhead. The Sun heats the Earth and, in turn, the Earth heats the air above it. The air rises, giving rainfall. The air spreads out, cools and sinks. This brings dry weather. The zone affected by these movements is the ITCZ. Each year, places in it have a brief rainy period and a longer dry period. The movement of the ITCZ varies with time and is unpredictable. There is, though, evidence that weather conditions within it have become more extreme	**Overall reduced rainfall.** As lands at the margins of deserts receive less rainfall they are incapable of supporting life and become desert	
	More intense storms. The rainfall is convectional. Thunderstorms have stronger winds and greater rainfall. Interception of heavy rain results in sheet erosion of the soil	
	More intense heating. A greater intensity of heat will cause greater evaporation. It will also reduce amounts available for people and their livestock and crops. Families are forced to migrate	

> **Revision activity**
>
> 1 Annotate a sketch of Figure 4 to show evidence of desertification.
> 2 Complete the table above with specific details of each 'effect' on a place or places that you have studied.

Human causes of desertification

Poor environment management is a major cause of desertification. However, it is not always intentional. Usually it is a result of lack of education or, more often, desperate action in order to survive.

Human cause	Specific place detail
Slash and burn of trees. This often occurs to clear land for commercial farming. Global multi-national corporations (MNCs) use the land to grow crops for consumption or for oils to convert into fuels. Such monoculture weakens the soil structure. It also takes land away from locals, creating more pressure on the remaining land	
Increased firewood demand. Most people in these areas use wood as a source of heating and cooking fuel. Fewer trees in an area and increased populations cause more destruction of trees, as naturally dying branches are too few to meet demand	
Less nomadism. There has been a move away from travelling with herds of animals in search pasture. This results in overgrazing of grass in those areas where the herders choose to settle. New grass is not allowed time to regenerate and dies	

Revision activity

1 Complete the following sequences to explain the effects of land management in areas experiencing desertification. The first one has been done for you:
Less interception of rain by plants → increased rainfall hitting soil → increased soil erosion → desertification.
Less tree shade for soil →
Unsustainable use of water from wells →
2 Complete the table above using detailed information from a place that you have studied.

Managing environments vulnerable to desertification

Managing desertification is often attempted at the local scale. However, desertification is a process that results in the extension of deserts into new areas of land in many linked countries. As such, like trans–boundary rivers, its management may be international in scale.

The Great Green Wall

REVISED

An agreement was signed in 2010 between eleven countries to grow a 15-km wide strip of trees and shrubs to attempt to prevent the southward spread of the Sahara Desert.

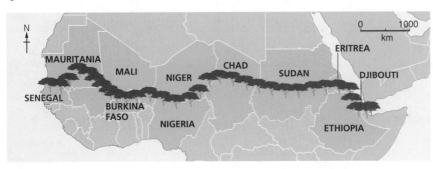

Figure 5 **The proposed location of the Great Green Wall.**

Possible benefits of the features of the Great Green Wall

Feature	Benefit
A Protects soil in rainy season	**1** so frees more time for other daily household tasks
B Leaves and other material decompose in soil	**2** so should be able to be managed by local people
C Provides accessible source of firewood for locals	**3** so reduces transpiration and increases yields
D Encourages farmers to grow fruit trees	**4** so reduces rates of soil erosion
E Provides shade for crops	**5** so reduces the chance of disputes over other issues
F Once established, relatively cheap to maintain	**6** so increases variety of diet, making it healthier
G Encourages cooperation between countries	**7** so adds to soil structure to resist erosion

How realistic is the Great Green Wall?

In two countries, Senegal and Niger, clear progress has been made. This has resulted in improved crop yields, better-fed livestock and increased local sources of firewood and medicines. In Senegal, locals are being encouraged to develop **ecotourism**. However, not a great deal has happened in the other nine countries. The wall is an example of **top-down development**. Local people are often suspicious of changes that are imposed on them from outside their area. This greatly slows progress.

Now test yourself

1 a) Work out the extent of the Great Green Wall.
  b) How might you change the map to make this calculation easier?
2 Link each of the features in the table with its correct benefit.

TESTED

Ecotourism Holidays aimed at having as little negative impact on an area as possible. The natural features of the area of destination are usually the attraction for the tourists

Top-down development Changes imposed on groups of local people by outside groups, often national governments

Local community changes: the bottom-up model

It is often difficult to encourage change. Communities have their own traditions that they feel have often served them well for centuries. Change also often comes at a cost they cannot afford. It is, though, possible for **non-government organisations (NGOs)**, usually charities based in HICs, to work closely with locals to implement change and to provide the funds required. Changes agreed and implemented at a local level are **bottom-up developments**.

> **Non-government organisations (NGOs)**
> Non-profit groups, usually charities based in HICs, that attempt to address a particular issue
>
> **Bottom-up development**
> Changes agreed and developed at the local level, actively involving the people who will be affected by them

Runoff is slowed by the bund, giving more time for infiltration

Rainwater infiltrates and recharges soil moisture

Bunds are placed 10–25 m apart

Any soil that has been eroded by runoff is trapped by the bund. Topsoil and organic matter (for example, leaf litter) are deposited here

Figure 6 How bunds, or 'magic stones' work.

Figure 7 Drip irrigation.

Drought-tolerant crops

Pearl millet is a very drought-tolerant crop. It provides food, fuel and construction material.

Plant breeders at the International Crops Research Institute for the Semi-Arid Tropics (ICRISAT) have produced new, improved and even more tolerant varieties.

Early maturing varieties of millet are very useful in helping semi-arid communities get through the 'hungry season'. This is the period before harvest, when the previous year's grain supplies have already been eaten.

Now test yourself

1. Look at the Figure 6, which shows the use of bunds in a semi-arid area. Write a paragraph to help explain how their use might help to prevent desertification.
2. Label Figure 7 to show how drip irrigation operates.
3. Annotate Figure 7 to help explain the advantages it will bring to farmers who use it and to the fight against desertification.

TESTED ☐

Problem solving

Unit 2: A different type of examination

The second unit of your three examinations is called 'Problem Solving Geography'. It is 1 hour and 30 minutes long and will have a focus on a place and problem to be solved somewhere in the world. The content is unlikely to be familiar to you – but don't worry, it's planned that way. Your examiners want to know how well you can apply your geographical abilities to a new situation.

Features of a problem-solving paper

Your examiners really want to know how *you* would solve a geographical problem. This exam paper will take you through a series of questions, finishing with the chance for you to say what you think and explain why you think it.

Whatever problem is put in front of you, it will always be an exercise in sustainability. You will be introduced to a current problem that exists somewhere in the world and asked how it should be managed for a sustainable future.

Every problem-solving paper will be in three parts:
● Part A: an introduction to a place and a problem to be solved
● Part B: an investigation of alternative possible solutions to the issue
● Part C: an opportunity to choose a solution or prioritise a list of options and justify your choice of strategy.

In general terms, the amount of challenge will increase as you work through the paper. There will be the greatest support for you in Part A. The amount by which the questions are broken down for you will be reduced in Part B. Part C will be a single question where you will have to provide your own structure.

Your mathematical abilities will be tested through this paper. You may be expected to apply your numerical and statistical skills to geography based on responses to resources like maps, graphs and tables.

Tackling the problem-solving experience

Over the next few pages you will be taken through a complete problem-solving experience. The nature of the questions is very similar to those on the other two units of your examination:
● Each question targets a particular Assessment Objective (AO) in the Geography Specification.
● Your examiners will be instructed to mark each question according to the AO being tested.
● Following each question you will be given both the number of marks available for answering the question and the AO that is being tested by the question. For example, '[4] [AO3]' means that the question has four marks available and that these are for Assessment Objective Three. Advice on how you should respond to questions targeting each AO is given on pages 10 and 11.

The problem-solving task

This is Part C of your problem-solving paper.

There is also a similar question to this on Unit 3, your Applied Fieldwork Enquiry paper. It comes in Part C and is called, 'The Wider UK dimension'. This question is similar to your problem-solving final task. It:

- offers alternative suggestions for future development in an area
- asks you to select and justify your views
- is accompanied by information in a Resource Folder
- is worth 12 marks for giving your opinions plus 4 for your written English abilities
- as on the problem-solving paper, the full 12 marks will be awarded for Assessment Objective Three.

Of course, unlike the problem-solving paper, which could be about anywhere, this question will focus on somewhere in the UK.

Planning for a problem-solving question

Planning: the key to success

- Carefully consider the options. Are you asked to make a straight choice or are you asked to prioritise those you have been given?
- Don't just use evidence that has been given to you. Incorporate other evidence that you have brought with you to the examination room from, perhaps, a place you have studied in class or even seen on TV news reports in order to back up your views.
- Now work out a structure for your report. It should include:
  - a clear statement of your chosen decision
  - an evaluation of the strategies that you have discarded or to which you have given a lower priority; make sure you place greatest emphasis on the disadvantages
  - an evaluation of the strategy you have chosen or to which you have given the highest priority; make sure you place greatest emphasis on the advantages
  - a final sentence briefly stating your choice.
- Ensure that your analysis demonstrates your knowledge and understanding of places and geographical principles.

A good plan is to remember the following:
- Use the simple statements that you have been given as starting points.
- Consider the social, economic and environmental implications of the choice you are making.
- Consider the short- and longer-term effects of your strategy. This is a major feature of sustainability.
- Write your report in a formal style and don't be tempted to waffle. You should need no more space for your response than the 1.5–2 pages provided for you in the question/answer booklet.
- Write in your best possible English. Four marks will be applied to this question for your ability to spell, punctuate and accurately use grammar and specialist terms.
- Remember to use specialist terms *effectively*. Integrate them into your account to help demonstrate your understanding as opposed to merely adding them to show that you know them.

Problem-solving paper

Part A

*You are advised to spend about **20 minutes** on this part. This part provides you with background information about Fort McMurray and the oil-extraction industry.*

(a) Study the map on page 1 of the separate Resource Folder. It shows the location of Fort McMurray.

(i) Name the continent in which Fort McMurray is located.

[1] [AO1] [1 line]

(ii) Fort McMurray is located in which province of Canada? Circle the correct answer.

Saskatchewan / Quebec / Alberta   [1] [AO4]

(iii) Give two more statements about the location of Fort McMurray.

[2] [AO4] [3 lines]

(b) Study Graph 1 on page 1 of the separate Resource Folder.

(i) Which Canadian source of oil was the most important in 2000?

[1] [AO4] [1 line]

(ii) Use information from Graph 1 to complete the statements below.

Between 2008 and 2020 it is expected that in Canada conventional oil and _____ will reduce production. _____ is expected to rise between these years to a total of _____ million barrels a day.

Show how you worked out this total.   [4] [AO4]

(iii) One way in which Canada may benefit from the sale of oil is by getting money in the form of taxation. Explain why this may affect people's quality of life.   [4] [AO2] [6 lines]

(c) There is concern that an increase in extraction of oil from oil sands will cause greater water pollution.

i) Label the photograph to show two other types of pollution that may be caused during the extraction of oil from oil sands.

[2] [AO1]

(ii) Describe how water pollution may affect the natural environment.

[5] [AO2] [8 lines]

[End of Part A: 20 marks]

Part B

*You are advised to spend about **30 minutes** on this part. This part examines the social, environmental and economic effects of extracting oil from sands in Canada to advise whether the industry should be stopped immediately, phased out slowly or allowed to fully continue.*

Some social effects of the oil extraction industry

(a) Study News Article 1 in the separate Resource Folder.

(i) Choose from the news article one disadvantage and one advantage of growing up in Fort McMurray. Explain your choice of each.

Disadvantage   [3 lines]

Advantage   [3 lines] [4] [AO2]

(ii) The growth of the oil industry has attracted a large number of migrants to Fort McMurray. Explain why this might be viewed with mixed feelings by the town's original community.

[5] [AO2] [9 lines]

(b) Study the graph below.

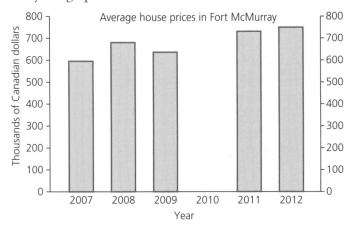

Since about 2010 the price of oil has fallen sharply. Large numbers of workers in the industry have been laid off.

(i) Complete the graph using the following information.
The average house price in Fort McMurray in 2010 was $675,000. [1] [AO4]

(ii) Use the information above to suggest why the changes shown may affect the housing market in Fort McMurray.

[4] [AO3] [8 lines]

Some environmental effects of the oil extraction industry

(c) Study the pie chart below.

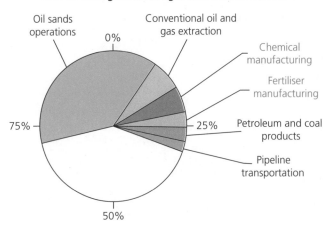

(i) Complete the pie chart to show the following information about greenhouse gas emissions.
37% came from electricity generation
4% came from other industries. [2] [AO4]

(ii) To what extent might the ending of Alberta's oil extraction impact on climate change? [4] [AO3]

(iii) Describe how climate change may affect people and environments. [4] [AO1] [7 lines]

(iv) Describe how greenhouse gases may contribute to climate change. [4] [AO1] [7 lines]

Some economic effects of the oil extraction industry

(d) Study Graph 2 in the separate Resource Folder.

    (i)   What were Alberta's oil reserves estimated to be in 2011?

                                                    [1] [AO4] [1 line]

    (ii)   Place the following areas of the world in rank order according to their oil reserves.

         Africa; The Middle East; North America; Russia.

                                                      [1] [AO1] [1 line]

    (iii) 'Extracting oil from oil sands will provide sustainable long-term employment opportunities in Fort McMurray.' Is this true? Use evidence from the graph to support your answer.

                                                      [4] [AO3] [7 lines]

    (iv) Describe how the multiplier effect operates in an area of 'boom and bust' fluctuating job opportunities.

                                                    [6] [AO2] [11 lines]

End of Part B: 40 marks

Part C

*You are advised to spend about **40 minutes** on this part. This part asks you to decide about the future of extraction of oil from the Alberta oil shales.*

The three options are:

(a) A complete ban on future oil extraction.

(b) A phasing out of extraction over a ten-year period.

(c) A continuation of extraction as long as it is economically profitable.

Write a letter to the government of Alberta in which you advise which of the three options it should adopt. In your letter you should:

• State which option you would advise the government to adopt.

• Justify your decision. You should take into account the social, economic and environmental sustainability of the options.

The Factfile in the separate Resource Folder gives further information about the issue.

Make use of information from this and other parts of the paper to help your answer. You may organise your ideas and plan your letter on the blank page that follows.

Your ability to spell, punctuate and use grammar and specialist terminology accurately will be assessed in your answer.

    To the government of Alberta.

    I am advising you to adopt the following option:

    _____

    I have decided on this option because ...

                                                           [12] [AO3] [2 pages]

End of Part C: 12 + 4 marks

End of examination: 72 + 4 = 76 marks

Resource Folder

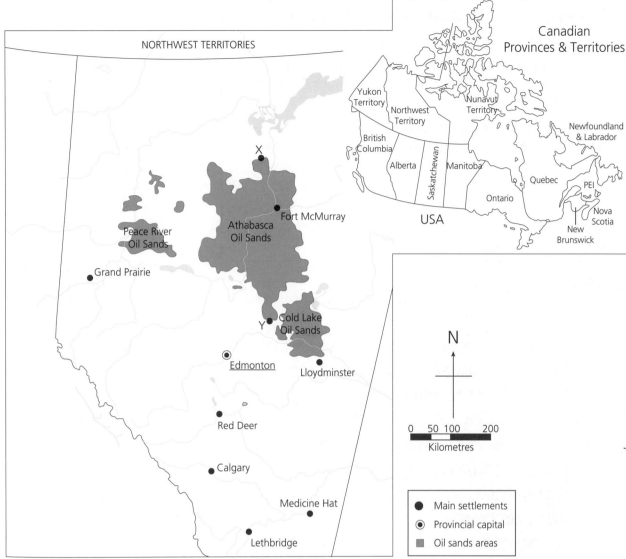

Map 1 Canadian oil sands areas, Alberta

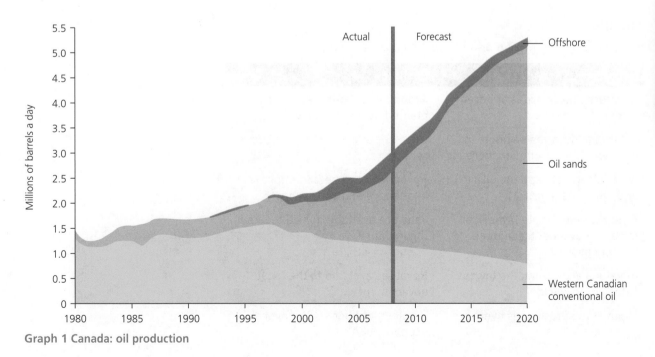

Graph 1 Canada: oil production

News article 1

The good and bad of boomtown life

Being a boomtown has its ups and downs.

On the upside, Fort McMurray offers excellent cultural experiences. Local theatre, arts, sports, music and other community events take place throughout the year.

On the other hand, there are problems of crime and homelessness. It is also now a 'shift-working city' with oil sands operations taking place 24 hours a day. This brings challenges for working, socialising and family relationships.

Living in a boomtown can be both exciting and challenging and changing prices of world oil markets means there are no guarantees for the future.

Graph 2

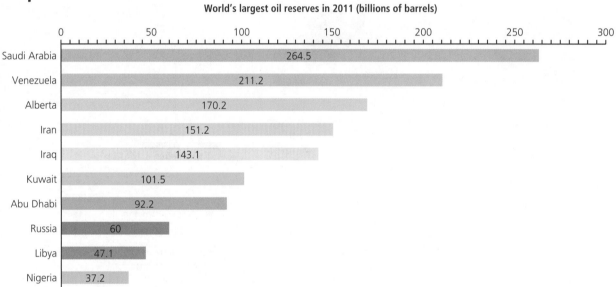

World's largest oil reserves in 2011 (billions of barrels)

Saudi Arabia	264.5
Venezuela	211.2
Alberta	170.2
Iran	151.2
Iraq	143.1
Kuwait	101.5
Abu Dhabi	92.2
Russia	60
Libya	47.1
Nigeria	37.2

Factfile

Some effects of extraction of oil from Alberta's oil sands:

Negative	Positive
It disturbs large areas of boreal forests	Around 8 million trees have been planted by the industry
Sewerage systems struggle to cope with the population increase	Over 90 per cent of water used is recycled
A wage gap between the oil-rich and other residents	Revenues have helped improve healthcare and education
First Nations people (American Indians) have lost traditional hunting grounds	Jobs are available for all locals in the industry
Royalties are among the lowest rates in the world	Royalties are paid to the local government
Much of the profit goes overseas to foreign MNCs	Employment boosts the overall economy of Alberta

Fieldwork

Unit 3: the Applied Fieldwork Enquiry

The first two parts of Unit 3 will be about the abilities that you have developed when taking part in two elements of field study with your school. During your field excursions and their follow-up in class you will have:

1 asked questions that help you to create an overall statement (hypothesis) to be tested by your fieldwork

2 collected data to help you answer these questions and test whether or not your hypothesis is true

3 processed and presented data using a variety of techniques to make it easier to understand

4 analysed your processed data to identify and to help explain patterns and trends

5 drawn conclusions about how closely your fieldwork findings match your hypothesis

6 then looked back at how well your investigation went. You will have evaluated the ways in which you collected data and the usefulness of the data itself, and worked out how you could have done it all better.

But what did you actually do?

The exam board changes the focus of fieldwork every year and the actual work you do will depend on the choices it makes for your particular exam year. These will be known when you start your course.

You will be tested on one of each of the following:

● **methodological approach:** the data that you collect to help you explore your hypothesis and the ways in which you collect it *and the*
● **conceptual framework:** a geographical concept or idea that you should already understand from your studies in geography before carrying out the fieldwork.

Revision activity

In the table below, tick the methodological approach and the conceptual framework needed for your exam.

Methodological approach	Tick one	Conceptual framework	Tick one
Use of transects For example, changes across an urban area, along a river channel, a beach		**Place** Understanding the identity of places. For example, a comparison of two coastal places, river features, villages or urban neighbourhoods	
Change over time For example, of land use, weather, coastal landforms		**Sphere of influence** Extent of catchment areas and effects of this on a place. For example, the effects of a sporting event, a honeypot site, a river catchment and its flood risk	
Qualitative surveys For example, of urban or natural environments or perceptions of flood risk		**Cycles and flows** Patterns of movement and reasons for or effects of them. For example, migration, commuter flows, river flows in two places at different times	
Geographical flows For example, of commuter movements, traffic flows, sediment along a coast		**Mitigating risk** The nature of risk and people's responses to it. For example, coastal/river flood risk strategies, climate change and local responses, environmental risk and its management	
		Sustainability How future-proof a place is and how it might be made more sustainable. For example, impact of transport schemes, development of a brownfield site, evaluation of flood prevention schemes	
		Inequality Analysing patterns of inequality. For example, access to services, quality of life for a particular group of people, urban regeneration and deprivation reduction	

You will use both the approach and framework you have ticked in two separate fieldwork enquiries. Through them you will explore both *human* and *physical* data.

My two enquiries

You are likely to be tested on a variety of aspects of your two enquiries. However, all of the testing will be based on examples and data provided by your examiners and not on the actual studies you did. So you will need to be able to transfer your fieldwork abilities to new situations in a similar way to the problem-solving skills we discussed earlier.

Revision activity

Complete each of the tables on pages 126 and 127 as an aid to revising what you did in each of your fieldwork enquiries. The divisions in the left-hand column are the stages named at the start of this chapter. Tick each of these stages only when you feel that you fully know and understand it and the processes it involved you in.

1 Write your overall hypothesis and any organising questions in the first two rows.
2 For each 'technique' space list the techniques and sources you used.
3 For each 'findings' space relate what you found out in relation to your hypothesis and sub-questions.
4 In the 'comments' spaces evaluate the techniques you used or state how closely your findings matched your overall hypothesis and its sub-questions.
5 Complete the 'improvements' space to list ways in which you would change the activity were you to do it again.

My first fieldwork enquiry

Hypothesis		
Sub-questions (tick when fully understood)		

	Technique	Comment
Primary evidence collection techniques		
Secondary evidence		
Processing and presentation techniques		

	Findings	
Analysis: patterns and trends		
Conclusions		
Improvements		

My second fieldwork enquiry

Hypothesis		
Sub-questions (tick when fully understood)		
Primary evidence collection techniques	**Technique**	**Comment**
Secondary evidence		
Processing and presentation techniques		
Analysis: patterns and trends	**Findings**	
Conclusions		
Improvements		